SpringerBriefs in Applied Sciences and Technology

SpringerBriefs present concise summaries of cutting-edge research and practical applications across a wide spectrum of fields. Featuring compact volumes of 50 to 125 pages, the series covers a range of content from professional to academic.

Typical publications can be:

- A timely report of state-of-the art methods
- An introduction to or a manual for the application of mathematical or computer techniques
- A bridge between new research results, as published in journal articles
- A snapshot of a hot or emerging topic
- An in-depth case study
- A presentation of core concepts that students must understand in order to make independent contributions

SpringerBriefs are characterized by fast, global electronic dissemination, standard publishing contracts, standardized manuscript preparation and formatting guidelines, and expedited production schedules.

On the one hand, **SpringerBriefs in Applied Sciences and Technology** are devoted to the publication of fundamentals and applications within the different classical engineering disciplines as well as in interdisciplinary fields that recently emerged between these areas. On the other hand, as the boundary separating fundamental research and applied technology is more and more dissolving, this series is particularly open to trans-disciplinary topics between fundamental science and engineering.

Indexed by EI-Compendex, SCOPUS and Springerlink.

More information about this series at https://link.springer.com/bookseries/8884

José Roberto Díaz-Reza ·
Jorge Luis García Alcaraz ·
Adrián Salvador Morales García

Best Practices in Lean Manufacturing

A Relational Analysis

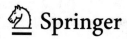

José Roberto Díaz-Reza 🆔
Universidad Autónoma de Ciudad Juárez
Ciudad Juárez, Chihuahua, Mexico

Jorge Luis García Alcaraz 🆔
Universidad Autónoma de Ciudad Juárez
Ciudad Juárez, Chihuahua, Mexico

Adrián Salvador Morales García 🆔
Universidad Autónoma de Ciudad Juárez
Ciudad Juárez, Chihuahua, Mexico

ISSN 2191-530X   ISSN 2191-5318   (electronic)
SpringerBriefs in Applied Sciences and Technology
ISBN 978-3-030-97751-1   ISBN 978-3-030-97752-8   (eBook)
https://doi.org/10.1007/978-3-030-97752-8

This Springer imprint is published by the registered company Springer Nature Switzerland AG
The registered company address is: Gewerbestrasse 11, 6330 Cham, Switzerland

Preface

Production and manufacturing processes are becoming increasingly complex, requiring complex control and management systems. Specifically, Mexico has made an effort to excel in the maquiladora industry sector. A maquiladora is a company that has its headquarters in another country and imports most of the raw materials, carries out assembly activities in the national territory, and then exports the finished products. This industrial-style takes advantage of the free trade agreements with the United States and Canada and the associated tariff benefits.

These manufacturing companies that come to Mexico bring with them many methods and machinery for their production processes, which are highly technological. One of these methodologies is lean manufacturing (LM), a set of tools focused on minimizing waste in production processes to minimize inventories, increase product quality and meet customer needs.

Although there are currently several studies focused on the maquiladora sector in Mexico and other countries, they are focused on analyzing some of the LM tools in isolation, so more integrative studies are needed, as LM practices are rarely implemented independently. This book aims to identify the level of implementation that these maquila industries have concerning LM best practices, but not by analyzing them independently. This book conducts a comprehensive and relational analysis since LM tools are not applied independently.

The importance of this industrial sector in Mexico is that currently (May 2021), there are 5185 companies of this type, 491 are established in the state of Chihuahua and, specifically, 328 are located in Ciudad Juárez. Economically speaking, from January to October 2020, maquiladora industry imports and exports were 201,099 and 213,656 million USD at the national level. At the state level, imports and exports were 22,184 22,637 million USD. Specifically, Ciudad Juárez had imports and exports of 16,638 and 16,978 million USD, respectively. These manufacturing companies generate 2,702,116 direct jobs, 486,057 in Chihuahua state, and 322,787 in Ciudad Juárez.

Given the economic and social importance of the maquiladora industry, this book focuses on this industrial sector and consists of seven chapters. Chapter one, called Lean manufacturing origins and concepts, as its name indicates, explains the concepts and origins of LM in the manufacturing industry. The chapter defines the main tools that integrate LM, identifies the most critical enablers and barriers for its implementation, reports some industrial successes implementation, indicates the benefits that companies are gaining, and finally, defines the research problem and objective of this book.

Chapter 2 describes some of the LM tools or best practices, their purpose in the industrial environment, applications, and expected benefits. Specifically, this book focuses on the following: Cellular layouts (CEL), Pull system (PUS), Small lot production (SLP), Quick setups (SMED), Uniform production level (UPL), Quality control (QUC), Total productive maintenance (TPM), Supply networks (SUN), Flexible resources (FLR) and Inventory minimization (INMI).

Chapter 3 defines the methodology followed in carrying out the study. The literature review is reported, the creation of the questionnaire and its application to the industrial sector, the data debugging, the techniques analyzed, and the conclusions obtained are briefly reported. Specifically, four models are reported in this book to relate the LM practices; the first three models report the variables or LM practices mentioned above. The fourth model reports a second-order model to integrate the different models into a single one.

Chapter 4 reports the first model, which is called Distribution and Maintenance and integrates four variables. The independent variables are Cell layout (CLA), Total productive maintenance (TPM), Single-Minute Exchange of Die (SMED), and Inventory minimization (INMI) is the dependent variable. Descriptive analysis for each variable is performed, six hypotheses are established, and conclusions are given based on the model and sensitivity analysis.

Chapter 5 presents model two, which is called Pull system and Quality control. It integrates four LM practices, the independent variables of which are Pull system (PUS), Small lot production (SLP) with six items, and Uniform production level (UPL). At the same time, Quality control (QUC) is the dependent variable. Six statistically tested hypotheses relate to the variables, and the conclusions and industrial implications from the relationships and sensitivity analysis are reported.

Chapter 6 reports model three, called Supplier network and Inventory minimization, and integrates four variables. The independent variables are Pull system (PUS), Small lot production (SLP), and Uniform production level (UPL), while the dependent variable is Quality control (QUC). The four variables are related using six hypotheses, which are statistically tested and concluded based on the results found from the model and the sensitivity analysis.

Finally, an integrative model with four second-order variables is reported since, as mentioned above, LM practices are not implemented in isolation in production systems. LM practices are integrated according to their role in the production process. The independent variables are Machinery (MAC) (Cellular layout, Total productive maintenance and Single-Minute Exchange of Die), Production Planning (PRP) (Push system, Small lot production, and Supplier networks), and Production process (PRR) (Inventory minimization, Uniform production level, and Flexible resources). In contrast, the dependent variable is Quality control (QUC).

Although this study and the models reported have been statistically validated with information from the Mexican manufacturing industry, this does not mean that they are not applicable in other geographical contexts. Hence, the authors hope that the results reported here will be helpful to all managers and engineers responsible for decision-making processes that seek to increase the efficiency of the production processes in their charge.

Ciudad Juárez, Mexico
José Roberto Díaz-Reza
Jorge Luis García Alcaraz
Adrián Salvador Morales García

Contents

List of Figures

List of Tables

Chapter 1
Lean Manufacturing Origins and Concepts

Abstract Lean Manufacturing is a systematic approach that identifies and eliminates waste in operations through continuous improvement, reducing the system's operating cost and meeting customer needs for maximum value at the lowest price. This chapter describes what lean manufacturing is, its tools, origins, facilitators, and barriers that prevent its implementation. It also presents applications made by some companies and the operational, social, and environmental benefits obtained from a correct implementation. Finally, the problem statement and the main objective of this research are described.

Keywords Lean manufacturing · LM house · LM origins · LM benefits · LM enablers

1.1 What is Lean Manufacturing?

Companies worldwide are trying to improve their profits without augmenting the selling price of their products, which can only be done by minimizing the cost of manufacturing products, increasing productivity, and reducing losses during the production process (Singh et al. 2018a, b). One of the philosophies that help achieve that is Lean Manufacturing (LM), an extension of just-in-time or Toyo-ta production system considered the primary comprehensive business strategy to improve performance (Womack et al. 1991).

LM is a revolutionary manufacturing philosophy compared to traditional standard mass production used for almost a century (Hosseini et al. 2015). LM practices enable producing a wider variety of products at a lower cost and higher quality while using fewer resources than traditional mass-production practices (Marodin et al. 2018).

The main goal of LM is to transform a company to be highly competitive and responsive to customer demand by eliminating waste (Bhamu and Sangwan 2014). Companies that implement LM become more effective by increasing product quality and value from the customer's perspective. Those companies become efficient by minimizing internal and external variability and reducing forms of waste in their information and production flows (Marodin et al. 2018).

Because of the above, LM is a systematic approach that identifies and eliminates waste in operations through continuous improvement, reducing the system's operating cost and meeting customer needs for maximum value at the lowest price (Abdulmalek and Rajgopal 2007). In that sense, LM stands for manufacturing without waste (Taj 2011).

Waste is anything other than the minimum amount of equipment, materials, parts, and labor time essential for the production and is of seven types: overproduction, lost lead time, transportation, inventory, processing, movement, and product defects (Taj 2008). Wyrwicka and Mrugalska (2017) mention a couple of additional wastes: new and unused talent and unsafe or ergonomic working conditions.

Thus, lean is a series of activities or solutions to eliminate waste, reduce non-value-added (NVA) operations and improve value-added (VA) (Wen et al. 2015) within the factory and along the supply chain (Panwar 2015). "Value" is any action that the customer would be willing to pay for Garre et al. (2017).

To achieve the above, LM adopts a holistic and multi-dimensional systems approach to understand and provide solutions to reduce waste, and thus develops close links between quality, cost, delivery, customer satisfaction, and continuous improvement (Kiatcharoenpol et al. 2015). LM is also an approach with the main objectives of developing knowledge and creating a working culture of continuous improvement to promote sustainability in-process operations and business management (Hallgren 2009).

Also, LM is an integrated manufacturing system to maximize capacity, reusability, and safety inventories by minimizing system variability (Salem et al. 2006). LM emphasizes the flow of materials from when a product begins to be manufactured until completed (Ruiz-De-Arbulo-Lopez et al. 2013).

1.1.1   The LM House

To achieve its objective, LM uses various tools and activities, sometimes grouped and called LM practices. Figure 1.1 illustrates the grouping of LM tools, which have been grouped into base tools that are fundamental and tools that are grouped into three pillars: tools that enable the flow of materials in the production system, tools that facilitate obtaining quality the first time, and finally, manufacturing system tools (Singh et al., 2018a, b).

The base tools are characterized by focusing on the diagnostics of the current state of the companies and serve as a starting point to be able to apply other tools. For example, quality cannot be achieved if the 5S tool has not been applied first or that the machines have an adequate maintenance program that avoids the production of defective parts for not having the calibration according to the technical specifications of the product (Colim et al. 2021; Raju et al. 2021).

Production flow tools allow reducing in-process inventories as possible, such as Just-in-Time (JIT), inventory management, one-piece flow, pull system, Kanban to improve material flow and facilitate production management. However, first and

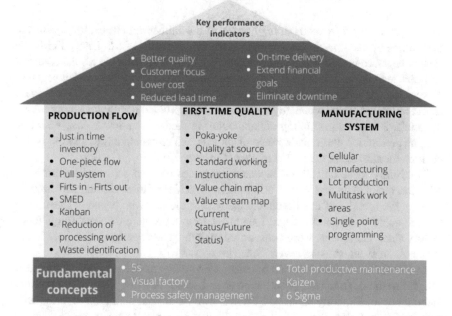

Fig. 1.1 Lean manufacturing tools

foremost, waste identification systems, which is one of the objectives of LM (Palange and Dhatrak 2021; Prada-Echevarría et al. 2021; Rahman et al. 2013).

Quality tools focus on ensuring that all product specifications are met and avoiding the generation of errors, such as poka-yoke, quality from the source, process value maps, standardized work instructions to operators by visual management, as well as education and training (Kurdve et al. 2019; Mulugeta 2020).

Finally, there are the tools focused or applied to manufacturing systems, such as layout or plant distribution, cellular manufacturing, group technology, batch production, with training to have multifunctional work areas that can perform different tasks and that the changes are fast (Annamalai et al. 2020; Dakov et al. 2010; Mohammadi and Forghani 2016).

The application of all these tools individually or grouped as manufacturing practices will allow the company to have a better product quality, better customer focus, generate products at lower cost and in a competitive manner, reduce delivery time, generate more cash flow for companies and eliminate waste, such as downtime on machines, tools, operators, and production lines in general.

1.2   Lean Manufacturing Origins

The origin of the LM methodology dates to the mid-twentieth century, with the beginnings of mass production by the automotive industry of Mr. Henry Ford. His system began to be adopted by other industries when observing the results obtained with this technique. The impact of its implementation in the global market began to be very high, and consequently, the Japanese position was affected, being relegated from the market. Thus, when Toyota began to have problems with the supply of raw materials, it focused on finding a manufacturing method at a lower cost, with low volume production and a wide range of models, which would allow them to better position themselves in the global market (Carreras 2010).

To achieve these objectives, Japanese industries focused on searching for optimization techniques that would allow them to achieve better results. Mr. Sakichi Toyoda, the founder of the Toyota company, was the one who began to create new tools to achieve this, and the first idea arose from a problem detected in his loom factory since the threads were constantly breaking. The operators were not always aware of the event, causing severe consequences.

Mr. Toyoda designed a device to detect when the yarns were breaking and send a signal to the operator to notice the error and solve it before producing a product that did not meet quality standards. Thus evolved a manual operation to one with a degree of automation, thus giving rise to the first lean manufacturing tool known as Jidoka (Sartal and Vázquez 2017).

When Mr. Toyoda analyzed the productivity and benefits obtained through Jidoka tool implementation, he began to look for new techniques to continue increasing the results. He began looking at production systems where people and machines would work in good synchrony and focused on achieving the same objectives, operating in a way that did not generate waste throughout the operations. That practice avoided producing units that would not be sent to the customer and focused on working according to demand and manufacturing only the parts necessary to complete production orders, which resulted in not generating overproduction and avoiding the consequences of doing so. Likewise, according to this new work technique's characteristics, it was called Just-in-Time (JIT) (Association 2018).

Continuing with the search to obtain better results and increase the company's productivity, they focused on the results obtained so far and the problems being detected. An opportunity for improvement was identified in machine tool changes since the operations required much time on some occasions and the clients asked to reduce the batches. We began to identify the total time from the last piece of the previous model to the first good piece of the next model. The activities were separated into two types, the internal ones that could only be done with the machine turned off and the external ones that could be done when the machine is operating. The main goal was to convert the internal to external and that the time did not exceed 10 min, this new technique was given the name of SMED (Single-Minute Exchange of Die) (Hemker et al. 2002).

Later, another LM founder, Mr. Taiichi Ohno, created another technique that allowed sending a signal to the stations to realize that it was necessary to produce new units, thus continuing with the ideology of only producing what was demanded. These signals are called Kanban, which moved along with the production, and in it came the general specifications of units to produce, continuing with this system. Later, with the support of Mr. Eiji Toyoda, they increased productivity and created the Toyota Production System (TPS) model (Rohlf 2015), giving greater support to the methodology designed by them, which is practical and recognized in the XXI century (Association 2018).

1.3   Enablers in LM Implementation

According to Singh (2014), LM encompasses five interdependent principles: (1) value definition, which involves knowing customer requirements, giving due weight or importance to those requirements and bringing them into product designs; (2) value stream analysis which is concerned with process design to add value to the product by customer requirements in the intended products; (3) flow, which requires production processes to be reliable so that the transformation of raw materials continues without interruption; (4) JIT or pull, which requires production to be carried out only when there is a confirmed customer order; and (5) perfection, which ensures that the four principles mentioned above are implemented perfectly.

A core concept of LM is pull production, where the flow on the factory floor is driven by the demand for production pulling down rather than traditional batch-based production, where production is im-pulsed from top to bottom according to a production schedule (Hallam and Contreras 2017). It can also be conceptualized as the strategy that generates internal flexibility to meet customer requests and eliminate waste in the production process (Koukoulaki 2014), so it frequently relies on machine maintenance (Zhou et al. 2019).

Successful LM implementation requires sophisticated, flexible, and numerically controlled machines; state-of-the-art technology, work systems, and procedures (Ghobakhloo and Azar 2018); standardization; flexible multi-skilled workforce, among others (Esmaeel et al. 2015; Singh, 2014). Likewise, for effective implementation of LM, management's commitment to increase working capital, provide staff training on lean philosophy, and develop mechanisms to reduce waste are crucial (Kafuku 2019).

LM requires striving for perfection by continuously eliminating waste discovered in the production process, thus requiring essential practice in product design, human resources, process and equipment, concurrent engineering, manufacturing planning, control, and supplier and customer relationship to consider the process (Kafuku 2019).

Salonitis and Tsinopoulos (2016) list some enablers and critical success factors in LM implementation, shown in Table 1.1.

Table 1.1 Enablers and critical success factors in LM implementation

Key drivers for implementing LM	Key success factor
To increase market share	Organizational culture and ownership
To increase flexibility	Developing organizational readiness
The need for survival from internal constraints	Management commitment and capability
Development of key performance indicators	Providing adequate resources to support change
Desire to employ world best practice	External support from consultants in the first instance
Part of the organization's continuous program	Effective communication and engagement
Drive to focus on customer	A strategic approach to improvements
Requirement/motivation by customers	Teamwork and joined-up whole systems thinking
Requirement by the mother company	Timing to set realist timescales for change and to make effective use of commitments and enthusiasm for change

Source Salonitis and Tsinopoulos (2016)

1.4   Barriers in LM Implementation

However, the introduction of LM in the manufacturing sector is not simple. It is not only a set of tools to be used but also a new management approach (also known as lean philosophy). So, it is necessary to consider several factors when establishing its implementation (Salonitis and Tsinopoulos 2016). In that sense, Nassereddine and Wehbe (2018) report a study conducted on 20 companies in the plastics industry in Lebanon, where they found that lack of planning and expanding customer requirements are the main barriers to LM implementation. Figure 1.2 summarily indicates the barriers reported by Salonitis and Tsinopoulos (2016).

In turn, Lodgaard et al. (2016) identify four categories of barriers to LM implementation. The first is associated with management and their lack of commitment, lack of leadership in lean projects, and lack of continuity. The second refers to the organization for LM, where roles are not clearly defined, lack of involvement and teamwork, and lack of motivation. The third refers to the tools and practices of LM, as often the proper techniques are not used or those that do not add value. Finally, the fourth category refers to their knowledge about the LM philosophy and their inability to obtain or disseminate it.

Recently, Sharma et al. (2020) report four barriers encountered in small and medium enterprises in India. They group into technical level barriers, organizational level barriers, cultural level barriers, and barriers that directly affect the industry's economics. As can be seen, the groupings of barriers are very similar.

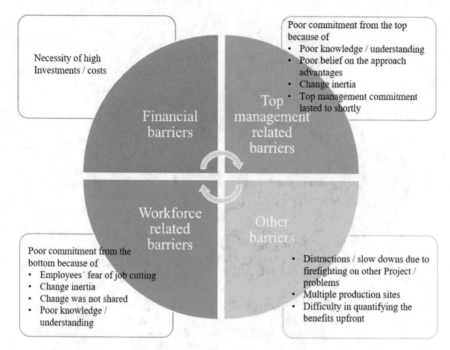

Fig. 1.2 LM barriers, according to Salonitis and Tsinopoulos (2016)

1.5 Lean Manufacturing: Some Applications

LM applications have spanned many industries, including automotive, electronics, home appliances, and consumer product manufacturing (Abdulmalek and Rajgopal 2007).

LM offers a variety of tools and techniques that assist companies in waste reduction. Tools such as value stream mapping (VSM), 5S methodology, jidoka, andon, Kanban, heijunka, among others (Kishimoto et al. 2020), JIT, cellular design, TPM, TQM empowered workforce (HRM) (Rahani and al-Ashraf 2012). A compendium of LM tools that are most discussed by academics can be seen in Fig. 1.3.

In their work, Miguez et al. (2018) integrated LM with people's expectations and the constant development of practices that guarantee people's safety, health, and quality of life. Such integration resulted in higher productivity and elimination, or reduction of waste related to movements or rework.

Ramakrishnan et al. (2019) implemented LM in a set of Indian companies, where they obtained significant quantitative and qualitative improvements. Among these improvements, they highlight the reduction of parts per million (PPM), the improvement of OEE (Overall Equipment Effectiveness), the improvement of 5S, the implementation of Kaizen, the reduction of model changeover time, and inventory reduction, and the improvement of productivity. Among the qualitative benefits, the improvement in employee morale, the importance of data collection for decision

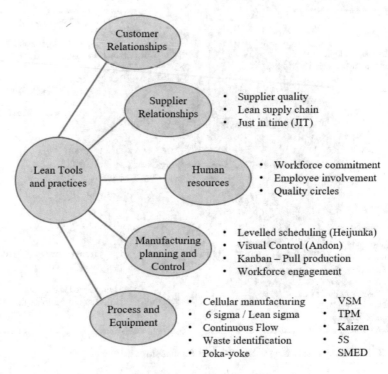

Fig. 1.3 LM tools and practices (Salonitis and Tsinopoulos 2016)

making, the knowledge of Pareto analysis, and the cause-and-effect diagram, among others, stand out.

Garre et al. (2017) used some LM tools to improve the welding process in the aerospace industry by implementing 5S redesigning the layout. The benefits obtained allow reducing transportation time and promoting continuous flow, decreasing the total operating cycle time.

Neves et al. (2018) implemented a combination of LM tools to identify problems and eliminate waste throughout the production process of small textile products. The tools implemented were PDCA, 5s, five whys, and two hows, and a significant impact was obtained in the fabric production process with gains of 10% in the valuable time available to the operator.

1.6   Lean Manufacturing Benefits

Companies implement LM to obtain operational, economic, social, and environmental benefits (Singh et al. 2018a, b). If one looks at the LM house, one of these benefits will be observed on the roof of the LM house.

Las empresas implementan LM para obtener beneficios, los cuales pueden ser operativos, económicos, sociales y ambientales Si se observa la casa de LM, en el techo de esta se observará alguno de esos tipos de beneficios.

1.6.1 Operational Benefits

LM practices improve manufacturing productivity by reducing setup times and in-process inventory work, improve production times, and thus improving market performance (Tu et al. 2006). LM creates a streamlined, high-quality system that produces products and services with higher productivity, lower cost, shorter lead times, and greater volume flexibility, ultimately improving organizations' performance (Shah and Ward 2003). By implementing LM, organizations could achieve process improvements dramatically impacting financial benefits, customer satisfaction, and production capacity (Kiatcharoenpol et al. 2015).

A list of operational benefits obtained in production lines that have applied LM is as follows: reduce costs (Keykavoussi and Ebrahimi 2020), implement improvement plans at all levels, define and implement in-plant indicators with a visual management system, implement TPM, SMED, and OEE (Prabowo and Adesta 2019), eliminate quality problems, reduce losses, scrap and rework (Purushothaman et al. 2020), and meet your delivery deadlines.

1.6.2 Social Benefits

One of the benefits of LM is that it offers greater safety to the workers who carry out activities on the production lines since they are the ones who handle the materials, many of which are high risk. That is why some authors indicate that LM is the basis for the social sustainability of the company (Salentijn et al. 2021). For example, Agrawal and Nath (2020) indicate that the agility and flexibility of the supply chain and LM practices support social sustainability since it requires high training, is multifunctional to have a continuous flow, and increases the levels of satisfaction in the operators.

Sahoo (2019) indicates that LM practices are indispensable to generate operational benefits in the first instance, which becomes social benefits when the human resources involved observe the improvements with their work, which increases motivation since the number of accidents is generally reduced. A particular case of the role of LM in social benefits is found in Pazos et al. (2009), where companies in the USA and Mexico are analyzed.

1.6.3 Environmental Benefits

One of the objectives of LM is to reduce waste. One of the most important ones refers to raw materials since mistakes can be made in the production process, so components must be discarded or reprocessed, which often requires spending energy on machines and paying specialized companies for their disintegration (Chen et al. 2020).

Kalyar et al. (2019) indicate that companies should try to comply with local and regional environmental rules and regulations; otherwise, economic sanctions and even suspension of work may occur. Likewise, Bai et al. (2019) indicate that when there are pollutant emissions in production processes, LM is a tool that helps improve companies' environmental performance. They should make investments to conform to the current regional regulations. More specifically, Alaa et al. (2017) state that there is a direct relationship between LM and the performance of companies so that tools focused on improving quality are the ones that strengthen this relationship the most.

1.6.4 Economic Benefits

The goal for a company is to generate financial profits for its partners, and LM is only the means to obtain them. The main economic benefits reported are an increase in the profitability of the company (Tretyakova et al. 2020), reduction of production costs by minimizing waste associated with raw materials, machine and tool downtime, as well as idle workers (Basu et al. 2020).

It has been shown that the relationship of LM with economic benefits is straight-forward (Chetthamrongchai and Jermsittiparsert 2019). To achieve it, it starts from selecting suitable suppliers that can provide raw materials in small batches and continuously. Hence, managers in production systems need to identify suitable customers that can support those fast flows (Ramadas et al. 2018).

1.7 Problem Research and Objective

Within the literature, works have been found where the effect of LM as an independent variable and the benefits it offers as independent variables are measured. For example, Möldner et al. (2018) measure LM's (technical and human practices) effect on the innovation process's performance in manufacturing organizations and organizational performance. Sahoo (2020) analyzes the benefits of implementing LM in the automotive industry in India, where the primary practices refer to the application of 5S, standardized work, and quality management practices, which is reflected in the higher productivity of the companies.

The above studies are interesting since they give a general idea of the relationship between the implementation of LM and their benefits. However, these studies tend to generalize the concept of LM when it is a set of tools. In that sense, other researchers have analyzed the impact of specific LM tools and the benefits obtained; for example, Vento et al. (2016) analyze Kaizen, García-Alcaraz et al. (2016), and Montes (2014) analyze JIT, Díaz-Reza et al. (2016) analyze SMED and Díaz-Reza et al. (2018) analyze TPM, among others. However, in industrial practice, no LM tool is applied in isolation, but instead, they are applied together, which is often referred to as LM best practices. Thus, these analyses with a single tool are reductionist, do not reflect the reality of the production systems, and studies focused on analyzing sets of tools are required.

This work aims to quantify the relationship between the primary LM practices and quality, one of the benefits most sought after by managers.

References

F.A. Abdulmalek, J. Rajgopal, Analyzing the benefits of lean manufacturing and value stream mapping via simulation: a process sector case study. Int. J. Prod. Econ. **107**(1), 223–236 (2007)

R. Agrawal, V. Nath, Agility and lean practices as antecedents of supply chain social sustainability. Int. J. Oper. Prod. Manag. **40**(10), 1589–1611 (2020). https://doi.org/10.1108/IJOPM-09-2019-0642

A.S. Alaa, A.I.M. Shaiful, M.Z. Zuraidah, A.M. Khalaf, in *The effect of lean manufacturing (LM) on environmental performance: a review study*, eds. by M.M.A. Abdullah, M.M. Ramli, S.Z. AbdRahim, S.S.M. Isa, M.N.M. Saad, R.C. Ismail, M.F. Ghazli. 3rd Electronic and Green Materials International Conference 2017, vol. 1885 (2017)

S. Annamalai, H. Vinoth Kumar, N. Bagathsingh, Analysis of lean manufacturing layout in a textile industry. Mater. Today: Proc. **33**, 3486–3490 (2020). https://doi.org/10.1016/j.matpr.2020.05.409

J.M. Association, *KANBAN: Y Just-in-time en Toyota* (Routledge, 2018)

C. Bai, A. Satir, J. Sarkis, Investing in lean manufacturing practices: an environmental and operational perspective. Int. J. Prod. Res. **57**(4), 1037–1051 (2019). https://doi.org/10.1080/00207543.2018.1498986

P. Basu, D. Chatterjee, I. Ghosh, P.K. Dan, Lean manufacturing implementation and performance: the role of economic volatility in an emerging economy. J. Manuf. Technol. Manag. (2020). https://doi.org/10.1108/JMTM-12-2019-0455

J. Bhamu, K.S. Sangwan, Lean manufacturing: literature review and research issues. Int. J. Oper. Prod. Manage. **34**(7), 876–940 (2014). https://doi.org/10.1108/IJOPM-08-2012-0315

M.R. Carreras, *Lean Manufacturing. La evidencia de una necesidad*: Ediciones Díaz de Santos (2010)

P.K. Chen, I. Lujan-Blanco, J. Fortuny-Santos, P. Ruiz-De-arbulo-lópez, Lean manufacturing and environmental sustainability: the effects of employee involvement, stakeholder pressure and iso 14001. Sustain. (Switzerland) **12**(18), 1–19 (2020). https://doi.org/10.3390/su12187258

P. Chetthamrongchai, K. Jermsittiparsert, Impact of lean manufacturing practices on financial performance of pharmaceutical sector in Thailand. Syst. Rev. Pharm. **10**(2), 208–217 (2019). https://doi.org/10.5530/srp.2019.2.29

A. Colim, R. Morgado, P. Carneiro, N. Costa, C. Faria, N. Sousa, P. Arezes, Lean manufacturing and ergonomics integration: defining productivity and wellbeing indicators in a human–robot workstation. Sustain. (Switzerland) **13**(4), 1–21 (2021). https://doi.org/10.3390/su13041931

I. Dakov, T. Lefterova, A. Petkova, Layout and production planning of virtual cellular manufacturing systems for mechanical machining. J. Econ. Asymmetries **7**(1), 43–67 (2010). https://doi.org/10. 1016/j.jeca.2010.01.004

J. Díaz-Reza, J. García-Alcaraz, L. Avelar-Sosa, J. Mendoza-Fong, J. Sáenz Diez-Muro, J. Blanco-Fernández, The role of managerial commitment and TPM implementation strategies in productivity benefits. Appl. Sci. **8**(7), 1153 (2018). https://doi.org/10.3390/app8071153

J.R. Díaz-Reza, J.L. García-Alcaraz, V. Martínez-Loya, J. Blanco-Fernández, E. Jiménez-Macías, L. Avelar-Sosa, The effect of SMED on benefits gained in maquiladora industry. Sustain. (Switzerland) **8**(12) (2016). https://doi.org/10.3390/su8121237

R.I. Esmaeel, I. Sukati, N.M. Jamal, The moderating role of advance manufacturing technology (AMT) on the relationship between large-supply chain and supply chain performance. Asian Soc. Sci. **11**(28), 37–44 (2015). https://doi.org/10.5539/ass.v11n28p37

J.L. García-Alcaraz, A.A.M. Macías, D.J.P. Luevano, J.B. Fernández, A.J.G. López, E.J. Macías, Main benefits obtained from a successful JIT implementation. Int. J. Adv. Manuf. Technol. **86**(9– 12), 2711–2722 (2016). https://doi.org/10.1007/s00170-016-8399-5

P. Garre, V.V.S. Nikhil Bharadwaj, P. Shiva Shashank, M. Harish, M. Sai Dheeraj, Applying lean in aerospace manufacturing. Mater. Today: Proc. **4**(8), 8439–8446 (2017). https://doi.org/10.1016/ j.matpr.2017.07.189

M. Ghobakhloo, A. Azar, Business excellence via advanced manufacturing technology and lean-agile manufacturing. J. Manuf. Technol. Manag. **29**(1), 2–24 (2018). https://doi.org/10.1108/ JMTM-03-2017-0049

C.R.A. Hallam, C. Contreras, The interrelation of lean and green manufacturing practices: a case of push or pull in implementation. Paper presented at the PICMET 2016—Proceedings of Portland international conference on management of engineering and technology: technology management for social innovation, 2017

H. Hemker, P. Giesen, R. AlDieri, V. Regnault, E. De Smed, R. Wagenvoord, S. Beguin, The calibrated automated thrombogram (CAT): a universal routine test for hyper-and hypocoagulability. Pathophysiol. Haemost. Thromb. **32**(5–6), 249–253 (2002)

A. Hosseini, H.A. Kishawy, H.M. Hussein, Lean manufacturing, in *Modern Manufacturing Engineering*. ed. by J.P. Davim (Springer International Publishing, Cham, 2015), pp. 249–269

J.M. Kafuku, Factors for effective implementation of lean manufacturing practice in selected industries in Tanzania. Proc. Manuf. **33**, 351–358 (2019). https://doi.org/10.1016/j.promfg.2019. 04.043

M.N. Kalyar, I. Shafique, A. Abid, Role of lean manufacturing and environmental management practices in eliciting environmental and financial performance: the contingent effect of institutional pressures. Environ. Sci. Pollut. Res. **26**(24), 24967–24978 (2019). https://doi.org/10.1007/ s11356-019-05729-3

A. Keykavoussi, A. Ebrahimi, Using fuzzy cost–time profile for effective implementation of lean programmes; SAIPA automotive manufacturer, case study. Total Qual. Manag. Bus. Excell. **31**(13–14), 1519–1543 (2020). https://doi.org/10.1080/14783363.2018.1490639

T. Kiatcharoenpol, T. Laosirihongthong, P. Chaiyawong, C. Glincha-em, A study of critical success factors and prioritization by using analysis hierarchy process in lean manufacturing implementation for Thai SMEs. Paper presented at the proceedings of the 5th international Asia conference on industrial engineering and management innovation (IEMI2014), Paris, 2015

K. Kishimoto, G. Medina, F. Sotelo, C. Raymundo, Application of lean manufacturing techniques to increase on-time deliveries: case study of a metalworking company with a make-to-order environment in Peru. Paper presented at the human interaction and emerging technologies, Cham, 2020

T. Koukoulaki, The impact of lean production on musculoskeletal and psychosocial risks: an examination of sociotechnical trends over 20 years. Appl. Ergon. **45**(2), 198–212 (2014)

M. Kurdve, U. Harlin, M. Hallin, C. Söderlund, M. Berglund, U. Florin, A. Landström, Designing visual management in manufacturing from a user perspective. Proc. CIRP **84**, 886–891 (2019). https://doi.org/10.1016/j.procir.2019.04.310

E. Lodgaard, J.A. Ingvaldsen, I. Gamme, S. Aschehoug, Barriers to lean implementation: perceptions of top managers, middle managers and workers. Proc. CIRP **57**, 595–600 (2016). https://doi.org/10.1016/j.procir.2016.11.103

G. Marodin, A.G. Frank, G.L. Tortorella, T. Netland, Lean product development and lean manufacturing: testing moderation effects. Int. J. Prod. Econ. **203**, 301–310 (2018). https://doi.org/10.1016/j.ijpe.2018.07.009

S.A. Miguez, J.F.A.G. Filho, J.E. Faustino, A.A. Gonçalves, A successful ergonomic solution based on lean manufacturing and participatory ergonomics. Paper presented at the advances in physical ergonomics and human factors, Cham, 2018

M. Mohammadi, K. Forghani, Designing cellular manufacturing systems considering S-shaped layout. Comput. Ind. Eng. **98**, 221–236 (2016). https://doi.org/10.1016/j.cie.2016.05.041

A.K. Möldner, J.A. Garza-Reyes, V. Kumar, Exploring lean manufacturing practices' influence on process innovation performance. J. Bus. Res. (2018). https://doi.org/10.1016/j.jbusres.2018.09.002

D Montes, Elements and benefits from JIT: a factor analysis. Master thesis, Universiddad Autónoma de Ciudad Juárez, Ciudad Juárez, Mexico, 2014

L. Mulugeta, Productivity improvement through lean manufacturing tools in Ethiopian garment manufacturing company. Paper presented at the Materials Today: Proceedings, 2020

A. Nassereddine, A. Wehbe, Competition and resilience: lean manufacturing in the plastic industry in Lebanon. Arab Econ. Bus. J. **13**(2), 179–189 (2018). https://doi.org/10.1016/j.aebj.2018.11.001

P. Neves, F.J.G. Silva, L.P. Ferreira, T. Pereira, A. Gouveia, C. Pimentel, Implementing lean tools in the manufacturing process of trimmings products. Proc. Manuf. **17**, 696–704 (2018). https://doi.org/10.1016/j.promfg.2018.10.119

A. Palange, P. Dhatrak, Lean manufacturing a vital tool to enhance productivity in manufacturing. Mater. Today: Proc. (2021). https://doi.org/10.1016/j.matpr.2020.12.193

A. Panwar, Lean implementation in Indian process industries—some empirical evidence. J. Manuf. Technol. Manag. **26**(1), 131–160 (2015). https://doi.org/10.1108/JMTM-05-2013-0049

P. Pazos, A.M. Canto, A. Powell, The impact of social system factors on sustainability of lean manufacturing: The case of the US and Mexico. Paper presented at the 30th Annual National Conference of the American Society for Engineering Management 2009, ASEM 2009

H.A. Prabowo, E.Y.T. Adesta, A study of total productive maintenance (TPM) and lean manufacturing tools and their impact on manufacturing performance. Int. J. Recent Technol. Eng. **7**(6), 39–43 (2019). Retrieved from https://www.scopus.com/inward/record.uri?eid=2-s2.0-85065190380&partnerID=40&md5=fe97f5965aafe0ddfa98e0c22ec1fa95

L. Prada-Echevarría, J. Chinchay-Grados, F. Maradiegue-Tuesta, C. Raymundo, Production control model using lean manufacturing tools and kanban/CONWIP systems to improve productivity in the process of sand casting in a heavy metalworking SME. In: Vol. 201. Smart Innovation, Systems and Technologies, pp. 439–447 (2021)

M.B. Purushothaman, J. Seadon, D. Moore, Waste reduction using lean tools in a multicultural environment. J. Clean. Prod. **265** (2020). https://doi.org/10.1016/j.jclepro.2020.121681

A.R. Rahani, M. al-Ashraf, Production flow analysis through value stream mapping: a lean manufacturing process case study. Proc. Eng. **41**, 1727–1734 (2012). https://doi.org/10.1016/j.proeng.2012.07.375

N.A.A. Rahman, S.M. Sharif, M.M. Esa, Lean manufacturing case study with Kanban system implementation. Proc. Econ. Financ. **7**, 174–180 (2013). https://doi.org/10.1016/S2212-5671(13)00232-3

G.S. Raju, M.K. Vanteru, A.C. Naik, Lean manufacturing for SMEs—A study with reference to SMEs. Paper presented at the AIP Conference Proceedings (2021)

T. Ramadas, K.P. Satish, K. Aibel Mathew, A model to identify the factors of supplier communication and financial availability to support the lean manufacturing implementation in small and medium scale enterprises. Int. J. Serv. Oper. Manage. **31**(4), 480–493 (2018). https://doi.org/10.1504/IJSOM.2018.096169

V. Ramakrishnan, J. Jayaprakash, C. Elanchezhian, B. Vijaya Ramnath, Implementation of lean manufacturing in Indian SMEs—a case study. Mater. Today: Proc. **16**, 1244–1250 (2019). https://doi.org/10.1016/j.matpr.2019.05.221

F.J. Rohlf, The tps series of software. Hystrix **26**(1) (2015)

P. Ruiz-De-Arbulo-Lopez, J. Fortuny-Santos, L. Cuatrecasas-Arbós, Lean manufacturing: costing the value stream. Ind. Manag. Data Syst. **113**(5), 647–668 (2013). https://doi.org/10.1108/026 35571311324124

S. Sahoo, Lean manufacturing practices and performance: the role of social and technical factors. Int. J. Quali. Reliab. Manage. **37**(5), 732–754 (2019). https://doi.org/10.1108/IJQRM-03-2019-0099

S. Sahoo, Assessing lean implementation and benefits within Indian automotive component manufacturing SMEs. Benchmarking **27**(3), 1042–1084 (2020). https://doi.org/10.1108/BIJ-07-2019-0299

O. Salem, J. Solomon, A. Genaidy, I. Minkarah, Lean construction: from theory to implementation. J. Manag. Eng. **22**(4), 168–175 (2006)

W. Salentijn, S. Beijer, J. Antony, Exploring the dark side of Lean: a systematic review of the lean factors that influence social outcomes. TQM J. (2021). https://doi.org/10.1108/TQM-09-2020-0218

K. Salonitis, C. Tsinopoulos, Drivers and barriers of lean implementation in the Greek manufacturing sector. Proc. CIRP **57**, 189–194 (2016). https://doi.org/10.1016/j.procir.2016.11.033

A. Sartal, X.H. Vázquez, Implementing information technologies and operational excellence: planning, emergence and randomness in the survival of adaptive manufacturing systems. J. Manuf. Syst. **45**, 1–16 (2017). https://doi.org/10.1016/j.jmsy.2017.07.007

R. Shah, P.T. Ward, Lean manufacturing: context, practice bundles, and performance. J. Oper. Manag. **21**(2), 129–149 (2003)

S.S. Sharma, P. Pandey, B.P. Sharma, Identification and categorization of lean manufacturing barriers in Indian SMEs. Paper presented at the AIP Conference Proceedings (2020)

J. Singh, H. Singh, G. Singh, Productivity improvement using lean manufacturing in manufacturing industry of Northern India: a case study. Int. J. Product. Perform. Manag. **67**(8), 1394–1415 (2018). https://doi.org/10.1108/IJPPM-02-2017-0037

S. Singh, S. Dixit, S. Sahai, A. Sao, Y. Kalonia, R. Subramanya Kumar, Key benefits of adopting lean manufacturing principles in Indian construction industry. Paper presented at the MATEC Web of Conferences (2018)

T. Singh, Role of manpower flexibility in lean manufacturing, in *The Flexible Enterprise* (Springer, 2014), pp. 309–319

S. Taj, Lean manufacturing performance in China: Assessment of 65 manufacturing plants. J. Manuf. Technol. Manag. **19**(2), 217–234 (2008). https://doi.org/10.1108/17410380810847927

S. Taj, The impact of lean operations on the Chinese manufacturing performance. J. Manuf. Technol. Manag. **22**(2), 223–240 (2011). https://doi.org/10.1108/17410381111102234

L.A. Tretyakova, M.V. Vladika, T.V. Tselyutina, T.A. Vlasova, O.A. Timokhina, Profitable production as a socio-economic based on supply chain management with lean production. Int. J. Supply Chain Manage. **9**(4), 1174–1181 (2020)

Q. Tu, M.A. Vonderembse, T. Ragu-Nathan, T.W. Sharkey, Absorptive capacity: enhancing the assimilation of time-based manufacturing practices. J. Oper. Manag. **24**(5), 692–710 (2006)

C.L. Wen, H.M. Wee, S. Wu, *Revisiting Lean Manufacturing Process with Vendor Managed Inventory System* (Paris, 2015)

J.P. Womack, D.T. Jones, D. Roos, M.I.o. Technology, *The Machine that Changed the World: The Story of Lean Production* (Harper Collins, 1991)

M.K. Wyrwicka, B. Mrugalska, Mirages of lean manufacturing in practice. Proc. Eng **182**, 780–785 (2017). https://doi.org/10.1016/j.proeng.2017.03.200

B. Zhou, G. Cheng, Z. Liu, Z. Liu, A preventive maintenance policy for a pull system with degradation and failures considering product quality. Proc. Inst. Mech. Eng., Part E: J. Process Mech. Eng. **233**(2), 335–347 (2019). https://doi.org/10.1177/0954408918784414

Chapter 2
Some Lean Manufacturing Tools

Abstract Lean manufacturing is a methodology that focuses on tools and practices implementation in production lines, which provides improvements and better performance in all operative processes. These tools optimize production processes to achieve cost reduction by reducing waste generated throughout the production process. This chapter gives an overview of 10 lean manufacturing tools used to optimize the manufacturing process, indicating the main benefits, enablers, and barriers.

Keywords LM description · LM tools

2.1 General Description of Lean Manufacturing

Lean manufacturing (LM) is a methodology that focuses on tools implementation, which provides improvements within their production processes. These new systems help identify activities that generate value to the production process and, likewise, those that do not (Belekoukias et al. 2014). In this way, LM allows the development of an improvement plan that allows streamlining the cycle time, eliminating tasks that are not essential in producing the good or service offered by the company (Yadav et al. 2020).

It should be emphasized that attention should be paid to those activities that do not generate value since sometimes they are not providing benefits. However, they cannot be eliminated because they are necessary to fulfill the production criteria (Jimenez et al. 2019). Those activities are commonly known as activities that do not generate value but are necessary.

Twenty-five tools make up lean manufacturing, which provides multiple competitive advantages to companies, obtaining benefits according to those implemented, since each has an objective to be fulfilled (Garza-Reyes et al. 2018). A group of them help to have a higher quality in the developed products, help reduce waste, and have higher utilization of the raw materials used (Venkat et al. 2020).

Other tools focus on the flow of materials, providing better use of the time and space available, helping to reduce the cycle times, and taking advantage of these

J. R. Díaz-Reza et al., *Best Practices in Lean Manufacturing*,
SpringerBriefs in Applied Sciences and Technology,
https://doi.org/10.1007/978-3-030-97752-8_2

resources in a better way (Kumar Banga et al. 2020). However, this chapter will only address the most critical tools shown in the following list:

- Cellular layouts (CEL)
- Pull system (PUS)
- Small-lot production (SLP)
- Quick setups (SMED)
- Uniform production level (UPL)
- Quality control (QUC)
- Total productive maintenance (TPM)
- Supply networks (SUN)
- Flexible resources (FLR)
- Inventory minimization (INMI)

This book describes their operation and their impact within the companies that adopt those LM tools, so it will be possible to observe the areas of improvement they support and the benefits obtained if they are correctly executed. The following is a brief description of each of these LM tools and practices.

2.2   Cellular Layouts (CEL)

The high competitiveness currently faced by companies generates that they are in continuous improvement and seek new ways to increase their productivity, relying on new technologies and methodologies that allow them to perform their activities more efficiently. To meet the quality requirements and delivery times currently demanded by customers, companies develop strategies that allow them to meet the needs of the market, taking into consideration the reduction of the cycle time of the good or service, being one of the points of most significant impact (Pantoja et al. 2017).

To meet the delivery times to customers, it is necessary to have lean planning regarding resources available in the company, the tools and machinery necessary for its development, and the labor required (Nallusamy 2016).

The Cell layout (CEL) tool has been adopted, which h to achieve positive acceptable delivery times CEL facilitates decreased cycle times, increased efficiency, and lower materials handling costs (Pattanaik and Sharma 2009).

One of the advantages of CEL integration is that it allows a better distribution of work areas, providing operators with greater ease in carrying out their activities (Arkat et al. 2012). CEL allows having all the necessary tools at a shorter distance and having materials and tools that are necessary for the completion of their tasks because they will only have what is strictly vital for their use, which allows a decrease in search time (Guzmán et al. 2018).

On the other hand, CEL allows operators to perform more than one activity at a time. Its distribution grants them the option of moving between worktables in a more agile way, which can be employed to delegate tasks, requiring fewer operators for task accomplishment (Kumar and Singh 2019). In addition, its non-linear design allows the same person who enters to start the production process to be the same person who terminates the cycle, thereby reducing the number of raw material and finished product storage locations. This situation is because some companies have loading vehicles responsible for supplying raw material and collecting the finished product. Thus, both sites for collection and storage would be located very close, which reduces distances and travel times (Kamaruddin et al. 2013; Lakshminarayanan et al. 2020).

Likewise, having CEL implemented helps to delimit the exclusive areas of cargo vehicles or forklifts as they are commonly known. For safety reasons, the areas through which they must circulate are delimited (Ham 1987). CEL allows companies to have a single route to drop off and pick up materials and products and greater control in this area (Gualtieri et al. 2020).

However, some authors declare that CEL also has some disadvantages when it is applied to production systems. Lietaert et al. (2019) indicate that CEL consumes huge space; several times, there is poor administration, ineffective communication from one cell to another, and expensive. Additionally, Lakshminarayanan et al. (2020) indicate that equipment cannot be shared and big lots are required, it is difficult to adjust the workspace to match workload, maintenance of standard operating procedure become difficult, and the training process in employees can be very long due to high specialization and skill level.

2.3 Pull System (PUS)

Nowadays, the high level of competition in the business market forces companies to adjust within the production systems to reduce costs and increase effectiveness. For this reason, production systems must adapt to the standards and levels that are planned in the goals, focusing on a decrease of waste and thus achieving a significant reduction of inventories. PUS prevents not manufacturing more than the company can sell, forcing it to store finished products and thus reducing the risks and costs of overproduction (Barón et al. 2012; Minh et al. 2019).

An important aspect to consider from PUS is that the demand behavior influences the production rate of the market. The market is the starting point for making production orders and planning the quantities each workstation must manufacture to complete the order (Jamaliah et al. 2016; Kimura and Terada 1981). In PUS requirements of the orders are relatively small.

To operate PUS in the manufacturing production process, another technique is implemented to help it indicate when to start producing and in what quantities, known as production Kanban cards. Kanban specifies the amount of product to be manufactured in each station, generating that only what is required is produced (Mahmood and Shevtshenko 2015; Puche et al. 2019).

One of the advantages of implementing PUS is that it provides greater control within the production processes, providing a technique that allows not producing more than what can be worked in each area. In PUS, the current station pulls the product from the previous station, working uniformly. PUS avoids delays and saturation of merchandise, which can cause accidents of a personal nature, not having enough space to perform their tasks, or damaging the product and impairing its quality and functionality.

One of the benefits of PUS adoption is that it helps companies eliminate one of the great wastes of overproduction and subsequent storage (Higuchi et al. 2015). PUS allows having better control of inventory levels, providing new advantages in the layout design. In this way, it will not be necessary to allocate too much space for the storage of the product, which will allow to use it for other areas that require it.

Other vital aspects must be considered by having a production designed under PUS since due to its design, it is necessary to have constant and updated communication with the sales department (Rusli et al. 2015). This sales behavior in the market lets to manufacture and complete the production orders (Lu et al. 2011). Similarly, constant communication with suppliers must be constant because the same amount of material is not always required. When a large order arrives, there must be the flexibility to increase the purchase orders of raw materials and supply them on time (Chopra and Meindl 2013).

2.4   Small-Lot Production (SLP)

Nowadays, non-mass production or small-batch production has become more critical within companies. This way of operating production processes allows greater flexibility, providing greater advantages to meet customer demand for special product orders (Zhang and Janet 2020). SLP allows companies to schedule according to the behavior of product demand, which reduces the risk of material being damaged by spending too much time in the warehouse waiting to be distributed (Müller and Sladojevic 2001).

On the other hand, other advantages that can be obtained with SLP are that the stocks of finished products of the different ranges that the company has are usually in small quantities, which means that the necessary facilities do not need to be very large (Sánchez et al. 2006). SLP provides greater control in the design and distribution of the work areas and reduces the risk of damage since the delivery time to the end customer is usually short. Likewise, the risk of obsolescence of the materials required for their manufacture is reduced (Alcaraz et al. 2014).

Under an SLP system, it must be taken into account for the success of the production objectives, to have a range of suppliers that are reliable with the quality of the material and with the delivery times of the orders, so as not to cause delays (Castro et al. 2009; De Castro Vila and Leal 2017). SLP does not require much raw material, which allows companies to have a smaller number of suppliers, which helps to have a better selection of them and only have a relationship with those who can meet their needs in terms of quality, quantity, and delivery time (Alcaraz et al. 2013).

Another benefit obtained by working with SLP is that it helps not to have an overproduction, being that the way of operating in mass has ceased to be effective over the years. Mass production is criticized due to the amount of capital needed to perform it and the process that is carried out that returns that investment is not usually a short time (Kazanskaia et al. 2015). Therefore, the company may have financial problems by not having the necessary income available to carry out the activities, maintenance and, if necessary, not having the budget to buy the raw material, so this small-batch technique allows attacking these problems (García et al. 2003).

2.5 Quick Setups (SMED)

Optimizing production processes and making them more agile is one of the most critical challenges and approaches since the loss of idle time brings high costs to companies. Therefore, searching for a tool to reduce these times is vital (Guzel and Asiabi 2020). The Single-Minute Exchange of Dies (SMED) technique ensures a quick and efficient changeover from a product produced to another to be produced (da Silva and Godinho Filho 2019). Model changeover time is the time taken from the last good part of one product to the first good part of the following product (Karam et al. 2018).

One of SMED's approaches focuses on keeping the machines running as long as possible, allowing production according to the installed capacity, widely used in the Just-in-Time (JIT) philosophy. JIT focuses on producing the quantities required when they are needed and only the quantities demanded (Kim and Shin 2019). To achieve delivery time goals, machinery needs to generate the set production, so SMED is the most effective method to carry out a production governed by the JIT system. It decreases the time needed to adjust and keeps operators available to perform other tasks (Shingo 2017).

One-way adjustments can be achieved faster is with operator training and implementing new technologies that allow the tasks to be performed. Training programs must be paramount when aiming to reduce these idle times. However, despite adopting new technology in the adjustments, the proposed solution will not have favorable results if workers are unaware of how they work and master these changes. On the contrary, it will bring consequences that cannot be predicted and considerably affect the planned production (Almomani et al. 2013; Díaz-Reza et al. 2017).

On the other hand, one way to prevent idle time in machines is to manage and generate skills in operators to recognize the internal SMED activities (activities that are performed only when the machine is stopped) and external ones (activities that are performed when the machine is running). In this way, several strategies can be aimed to transfer these internal activities to external ones, thus significantly reducing downtime and converting it into adequate time. However, the necessary care must be taken to ensure that these changes are made efficiently to have the planned quality (Ulutas 2011).

2.6   Uniform Production Level (UPL)

Uniform production level (UPL) is an LM practice that aims to reduce production-level variability caused by changes in customer demand (Nawanir et al. 2016). UPL requires materials to pass through the production line in a pattern of uniform loads to reduce variation in variety and quantity over time (White et al. 1999).

Uniform workload stabilizes and smoothest production by reducing fluctuations in daily workload (White Richard and Pearson John 2001). UPL is achieved by producing the same combination of end products every day and only those products that will be sold or are already sold under contract (White Richard and Pearson John 2001).

Uniform workloads can reduce resource waste, increase capacity utilization (Hamzeh et al. 2019), generate consistent production, increase flexibility, and support PUS implementation. It also allows for a predictable process and greater product flow throughout the processes with minimal inventory (Nawanir et al. 2018). The concept of uniform workload extends beyond operations activities to all system activities; therefore, it provides stability to the system (White Richard and Pearson John 2001).

2.7   Quality Control (QC)

Quality control (QC) can be defined as "part of quality management focused on meeting quality requirements" (ASQ 2021). QC aims to ensure that product quality is maintained and improved through product inspection and process control, which can help manufacturers meet consumer demand and reduce a quality loss (Chen and Tirupati 1995).

According to Montgomery (2020), the quality of a product can be evaluated along several dimensions: performance, reliability, durability, serviceability, aesthetics, features, perceived quality, and conformance to specifications or standards. QC evaluates that incoming raw materials, containers, labels, in-process materials, and finished products are fit for use. It also evaluates process performance against established standards and limits and determines lot acceptance criteria before distribution (Jani 2016).

According to Zhao et al. (2011), the first thing to do in any QC program is to identify the product's functional characteristics and nominal values and tolerances. Then, formulate a quality program that incorporates statistical process control tools to measure variable and attribute data for product conformance characteristics. ASQ recommends mastering the seven essential tools to ensure quality control: cause-and-effect diagram, check sheet, control charts, histograms, Pareto charts, scatter diagrams, and stratification (ASQ 2021).

2.8 Total Productive Maintenance (TPM)

TPM is a methodology originating from Japan to support its LM system, as reliable and effective equipment is essential for implementing LM initiatives in organizations (Arai and Sekine 1998). With TPM, companies can improve productivity and efficiency in maintenance activities (Mishra et al. 2021). TPM is a systematic program focused on production improvement that addresses the reliability of company facilities and the successful organization of plant resources through continuous employee involvement and empowerment in manufacturing, maintenance, and industrial function (Ahuja and Khamba 2008).

The objectives of TPM are to have machines with zero breakdowns, no slow running, no defects, and to make the production environment safe and in perfect condition (Salonitis and Tsinopoulos 2016). According to Nakajima (1988), TPM activities eliminate the six major losses: equipment failure, setup, adjustment time, idling and minor stoppages, reduced speed, process defects, and reduced yield.

TPM promotes autonomous operator maintenance through day-to-day activities that involve the entire workforce and increased employee morale and job satisfaction (Singh et al. 2013). These objectives require strong management support and the continuous use of work teams and small group activities to achieve incremental improvements (Cooke 2000).

According to Kiran (2017), TPM starts with cleaning, as the simple job of cleaning transforms into high-quality standards for a company. Cleaning allows inspecting machinery, which reveals abnormalities, which identifies abnormalities that allow for rectification, resulting in improvement that brings positive results that increase high-quality standards. Finally, these quality standards give pride to the work environment.

Pinto et al. (2020) mention that the way to implement and organize TPM is as follows:

- To eliminate the main problems, promoting the analysis and their causes to be eliminated and reduced.
- Autonomous maintenance implementation consists of giving responsibility to workers to perform maintenance routines.
- The planned maintenance program consists of systematic planning of maintenance activities by qualified technicians or by the worker himself.
- Education and training programs provide workers and operational managers with theoretical and practical information about machines to avoid losses.

The correct implementation of TPM brings with it a series of both tangible and intangible benefits. For example, Chan et al. (2005) report tangible benefits in quantitative form, average unit among attendances (replacing OEE), number of improvements, one-point lesson, suggestions, and training hours skill level. As for intangible benefits, the authors report gaining recognition and acceptance of individual responsibility for the team, development of a "can do" attitude and ownership for autonomous maintenance members, establishing a sense of importance for maintaining primary team conditions, development of problem-solving skills for team members, the concept of developing quality maintenance in production, and establishing cross-functional teamwork in production.

However, Attri et al. (2013) and Alseiari et al. (2020) report some technical and operational barriers for a successful TPM implementation as operations (production) management does not own OEE, the team is not educated on the whats and whys of TPM in the company, failure to start with operator-involved maintenance, inspections get too technical and too complicated too fast, TPM deployment is superficial, inspection results and proactive work identification goes nowhere and misalignment in the rewards structure.

2.9 Supplier Networks (SUN)

A buyer–supplier network is a complex network consisting of numerous buyers and suppliers that interact with each other in a "non-simple" way (Simon 1972). Campbell (1997) defines a buyer–seller network as a series of triadic relationships designed to generate value for the customer and build a sustainable competitive advantage for the creator and manager. These networks describe the cooperation between independent companies along a value chain to create a strategic advantage for the whole group that uses the core capabilities of the other members to create value for the end customer and thus create a competitive advantage.

Therefore, it is vital to evaluate and select suppliers, and this is an essential task in supply chain management (Li and Dong 2015). Such supplier selection allows finding the most suitable suppliers, who can provide the buyer with the right quality products and services at the right price, in the right quantities, and at the right time (Memari et al. 2013).

Relationships formed with buyers and suppliers are of particular importance due to their long-term nature and repeated interactions involving multiple firms, leading to increased commitment, trust-building, and knowledge accumulation from both parties (Gosling et al. 2010; Kotabe et al. 2003).

Cooperation with suppliers is critical when manufacturing companies change orders unexpectedly or request supplies in a shorter time than suppliers (He et al. 2014; Noordewier et al. 1990). Many producers want to build effective relationships with suppliers to improve their performance and competitive ability (Aghajani and Ahmadpour 2011), as the management of these relationships play a decisive role in the firm's ability to compete and profitability (Yan et al. 2015).

An important part of supplier relationship management is supplier development to increase the supplier's performance and capabilities and meet the buying firm's supply needs (Govindan et al. 2010). For supplier development, two-way communication, top management involvement, purchasing teams for a relatively large percentage of the supplier's output are important (Krause 1999; Tran et al. 2021).

According to Govindan et al. (2010) and Saghiri and Mirzabeiki (2021), the objective of supplier development can range from corrective to strategic. Corrective can be training a supplier's personnel in statistical process control to help them achieve desired quality levels and strategic when the buyer has competitive priorities that dramatic improvements in supplier capabilities can only meet.

Vendor-managed inventory (VMI) is a supply chain (SC) management approach to improve the SC performance of multiple companies while establishing a mutually beneficial relationship between a supplier and a retailer (Dasaklis and Casino 2019). In a VMI process, the company assigns to a supplier the task of managing the sequential link in its production chain, determining when and how much of each product should be shipped to its immediate customer. Currently, VMI is one of the most discussed collaborative practices to improve SC efficiency (Freitas et al. 2013).

The idea revolves around the manufacturer directly managing inventory at the customer (Whipple and Russell 2007). VMI is a supplier-managed inventory initiative, where the supplier assumes responsibility for planning and managing the customer's inventory based on an agreed-upon replenishment service contract (Formigoni et al. 2020). With the popularity of VMI, more and more suppliers are starting to manage their inventory at the customer's location, where suppliers are not paid for their delivery inventory until their goods are used (Wei 2020).

In VMI, the supplier decides the appropriate inventory levels of each product and the appropriate inventory policies to maintain those levels. The retailer provides the supplier with quick and real-time access to their inventory level (Sari 2007).

VMI effectively mitigates the bullwhip effect and offers two possible sources of reducing the bullwhip effect (Wen et al. 2015). First, decision making is reduced due to direct contact. Second, waiting time is reduced by reducing delays in information flow (Disney and Towill 2003).

The adoption of VMI approaches can have significant benefits for retailers, salespeople, and the SC as a whole (Sari 2007). For retailers, VMI provides a more efficient framework for order processing while reducing operational and administrative costs. VMI provides the visibility needed for better demand forecasting and, thus, more accurate inventory management (Dasaklis and Casino 2019).

However, it must be said that not all VMI implementations are successful, as there are prerequisites to implement them successfully. Before implementing a VMI model, there are strategic and operational critical success factors (Dasaklis and Casino 2019).

2.10 Flexible Resources (FLR)

Successful LM implementation requires a flexible, multi-skilled workforce and sophisticated numerically controlled and flexible machines, among other things (Singh 2014). A multi-skilled workforce allows responding quickly to unexpected and unbalanced demands that may appear (Chauhan and Singh 2013). When companies pursue a high division of labor and foster specialized skills in their workers, they will be less flexible than firms that rely on a more broadly educated worker who can adapt quickly to new products, engineering changes, or new technologies (Chauhan 2016).

Work flexibility comprises multiple skills, willingness to change and improve, and a flexible attitude toward new products, processes, and customers (Chauhan and Singh 2011). Workers can multi-task and be quickly assigned to different machine stations and are critical to the rapid adaptations required by today's stochastic manufacturing environments (Karuppan 2004). Such labor flexibility has been recognized as a tool for improving manufacturing performance through waste and resource reduction, which helps companies get closer to implementing LM (Kennedy 2003; Singh 2008).

According to Chen et al. (1992), labor flexibility is the ability of labor to perform a wide range of manufacturing tasks efficiently. Tsourveloudis and Phillis (1998) define labor flexibility as the ease of moving personnel to different departments within an organization and is achieved through the ability of multi-skilled personnel to perform a wide variety of tasks.

Such flexibility can also be exported to machines and tools in production lines because if these are flexible, they play an essential role in ensuring LM's success and generating a good proposition for the organization's long-term success (Chauhan and Singh 2013). Highly automated but flexible machines can produce complex product mixes with short setup times, lower inventories, and virtually no breakdowns. In that sense, machine flexibility is an essential requirement for implementing LM. Therefore, resource flexibility (labor and machinery) is essential for LM implementation (Chauhan 2016).

2.11   Inventory Minimization (INMI)

SC management is one of the most critical areas of research aimed at improving overall SC performance. Companies in these SC continuously seek to minimize their inventory levels to obtain cost savings over time (Chang 2007; Tliche et al. 2020). Some types of inventories are managed within companies, e.g., office equipment inventories, raw material inventories, work-in-process inventories, and finished goods inventories.

In manufacturing systems, random occurrences of machine failures cause disruptions in production. To reduce the impact of machine failures, buffers are implemented between machines, which increases the level of work-in-process inventory (WIP), increasing the operating cost (Ma and Koren 2004).

WIP inventory is measured by the number of items in each buffer and should remain small because inventories remain in the factory and do not generate profits. High inventories increase cycle time and decrease customer responsiveness, requiring more space and expensive equipment handling, increasing capital investment. Inventory quality decreases as unfinished items remain in the factory because they are vulnerable to damage (Conway et al. 1988; Gershwin 1994).

Limiting the amount of WIP reduces warehousing, financial cost and allows quality problems to be identified quickly. Also, it enables rapid response to machine breakdowns, material shortages, or worker unavailability (Hopp and Spearman

2011). In industrial practice, production planners use the WIP inventory level profile to control material flow and simplify production control (Lin et al. 2009; Susanto 2018).

WIP control is a proven approach to improving manufacturing systems and services (Qiu 2005). The benefits of WIP are that it prevents bottleneck machines from running out of material to process, reduces their idle time, and maintains high resource utilization (Sürie and Reuter 2015).

Another critical factor of SCs is raw material procurement, as an efficient decision could generate higher revenue growth (Taleizadeh and Noori-daryan 2016). Higher raw material inventory leads to an increase in total inventory costs, and managers should consider the ordering cost and holding cost, so the total inventory cost does not increase (Susanto 2018). Optimizing raw material inventory for the company reduces the total cost and improves the non-fulfillment of customer orders (Chen et al. 2017).

The above paragraph shows that raw material inventory is an essential factor in the production process. If the production process faces a shortage of raw material supply, it could be interrupted. However, inventory costs may increase if inventory is too large (Mulyana and Zuliana 2019). In addition, excess inventory can tie up capital and lead to waste as materials become obsolete (Chandra and Tulley 2016). Therefore, those who manage raw materials must control raw materials well so that there are no shortages or excess raw materials. When there are shortages, they can be overcome and produced smoothly (GS et al. 2018).

Finished goods inventory is the number of manufactured products on hand waiting to be sold to the customer (Paul and Azeem 2011). Manufacturers must maintain an optimal quantity to reduce costs and maximize SC efficiency. It aims to supply the required quantity of finished goods at the right place, at the right time, and at low cost (Sunil Kumar et al. 2018).

A critical operations management challenge facing manufacturing and distribution companies is managing their inventory and, more importantly, the level of inventory turnover (Andreou et al. 2016). Inventory turnover is the cycle of use and replacement of goods (Reynolds 1999), the ratio of a company's cost of goods sold to its average inventory level. It is commonly used to measure the performance of inventory managers, compare inventory productivity among retailers, and evaluate performance improvements over time (Gaur et al. 2005). Higher inventory turnover means that the company must invest less capital in WIP raw materials or finished goods (Demeter and Matyusz 2011).

References

H. Aghajani, M. Ahmadpour, Application of fuzzy topsis for ranking suppliers of supply chain in automobile manufacturing companies in Iran. Fuzzy Inf. Eng. 3(4), 433–444 (2011). https://doi.org/10.1007/s12543-011-0096-3

I. Ahuja, J. Khamba, An evaluation of TPM initiatives in Indian industry for enhanced manufacturing performance. Int. J. Qual. Reliab. Manage. (2008)

J.L.G. Alcaraz, A.A. Iniesta, A.A.M. Macías, Selección de proveedores basada en análisis dimensional. Contaduría y Administración **58**(3), 249–278 (2013)

J.L.G. Alcaraz, A.A. Maldonado, A.A. Iniesta, G.C. Robles, G.A. Hernández, A systematic review/survey for JIT implementation: Mexican maquiladoras as case study. Comput. Ind. **65**(4), 761–773 (2014). https://doi.org/10.1016/j.compind.2014.02.013

M.A. Almomani, M. Aladeemy, A. Abdelhadi, A. Mumani, A proposed approach for setup time reduction through integrating conventional SMED method with multiple criteria decision-making techniques. Comput. Ind. Eng. **66**(2), 461–469 (2013)

A.Y. Alseiari, P. Farrell, Y. Osman, *Technical and Operational Barriers that Affect the Successful Total Productive Maintenance (TPM) Implementation: Case Studies of Abu Dhabi Power Industry*, in Vol. 166. 32nd International Congress and Exhibition on Condition Monitoring and Diagnostic Engineering Management, COMADEM 2019 (Springer Science and Business Media Deutschland GmbH), pp. 1331–1344

P.C. Andreou, C. Louca, P.M. Panayides, The impact of vertical integration on inventory turnover and operating performance. Int. J. Log. Res. Appl. **19**(3), 218–238 (2016). https://doi.org/10.1080/13675567.2015.1070815

K. Arai, K. Sekine, *TPM for the Lean Factory: Innovative Methods and Worksheets for Equipment Management* (CRC Press, 1998)

J. Arkat, M.H. Farahani, F. Ahmadizar, Multi-objective genetic algorithm for cell formation problem considering cellular layout and operations scheduling. Int. J. Comput. Integr. Manuf. **25**(7), 625–635 (2012)

A.S.f.Q. ASQ, Learn about quality (2021). Retrieved from https://asq.org/quality-resources/quality-assurance-vs-control

R. Attri, S. Grover, N. Dev, D. Kumar, Analysis of barriers of total productive maintenance (TPM). Int. J. Syst. Assur. Eng. Manage. **4**(4), 365–377 (2013). https://doi.org/10.1007/s13198-012-0122-9

A.M. Barón, J.T. López, J.A.S. Mejía, Comparación y análisis de algunos sistemas de control de la producción tipo" pull", mediante simulación. Scientia Et Technica **17**(51), 100–106 (2012)

I. Belekoukias, J.A. Garza-Reyes, V. Kumar, The impact of lean methods and tools on the operational performance of manufacturing organisations. Int. J. Prod. Res. **52**(18), 5346–5366 (2014)

A.J. Campbell, Using buyer-supplier networks to increase innovation speed: an exploratory study of thai textile exporters. Asia Pacific J. Manage. **14**(2), 107–122 (1997). https://doi.org/10.1023/A:1015441315710

W.A.S. Castro, Ó.D.C. Gómez, L.F.O. Franco, Selección de proveedores: una aproximación al estado del arte. Cuadernos De Administración **22**(38), 145–167 (2009)

F.T.S. Chan, H.C.W. Lau, R.W.L. Ip, H.K. Chan, S. Kong, Implementation of total productive maintenance: a case study. Int. J. Prod. Econ. **95**(1), 71–94 (2005). https://doi.org/10.1016/j.ijpe.2003.10.021

V. Chandra, M. Tulley, *Raw Material Inventory Strategy for Make-To-Order Manufacturing* (Massachusetts Institute of Technology, 2016)

G. Chang, Analysis of inventory level under procurement constraints in supply chain. Front. Mech. Eng. China **2**(3), 361–363 (2007). https://doi.org/10.1007/s11465-007-0063-1

G. Chauhan, An analysis of the status of resource flexibility and lean manufacturing in a textile machinery manufacturing company. Int. J. Organ. Anal. (2016)

G. Chauhan, T.P. Singh, Lean manufacturing through management of labor and machine flexibility: a comprehensive review. Glob. J. Flex. Syst. Manag. **12**(1), 59–80 (2011). https://doi.org/10.1007/BF03396599

G. Chauhan, T.P. Singh, Resource flexibility for lean manufacturing: SAP-LAP analysis of a case study. Int. J. Lean Six Sigma **4**(4), 370–388 (2013). https://doi.org/10.1108/IJLSS-10-2012-0010

G.-H. Chen, Y. Zhao, B. Su, Raw material inventory optimization for MTO enterprises under price fluctuations. J. Discrete Mathe. Sci. Crypt. **20**(1), 255–270 (2017). https://doi.org/10.1080/097 20529.2016.1178930

I. Chen, R. Calantone, C. Chung, The marketing-manufacturing interface and manufacturing flexibility. Omega **20**(4), 431–443 (1992)

W.-H. Chen, D. Tirupati, On-line quality management: integration of product inspection and process control. Prod. Oper. Manag. **4**(3), 242–262 (1995). https://doi.org/10.1111/j.1937-5956.1995.tb0 0055.x

S. Chopra, P. Meindl, *Administración de la cadena de suministro* (Pearson educación, 2013)

R. Conway, W. Maxwell, J.O. McClain, L.J. Thomas, The role of work-in-process inventory in serial production lines. Oper. Res. **36**(2), 229–241 (1988)

F.L. Cooke, Implementing TPM in plant maintenance: some organisational barriers. Int. J. Qual. Reliab. Manage.

I.B. da Silva, M. Godinho Filho, Single-minute exchange of die (SMED): a state-of-the-art literature review. Int. J. Adv. Manuf. Technol. **102**(9), 4289–4307 (2019). https://doi.org/10.1007/s00170-019-03484-w

T. Dasaklis, F. Casino, Improving vendor-managed inventory strategy based on internet of things (IoT) applications and blockchain technology. Paper presented at the 2019 IEEE International Conference on Blockchain and Cryptocurrency (ICBC), 14–17 May 2019

R. De Castro Vila, G.G. Leal, Formando a Lean Thinkers. Direccion Y Organizacion **62**, 47–54 (2017). Retrieved from https://www.scopus.com/inward/record.uri?eid=2-s2.0-85032958821& partnerID=40&md5=4d7a0e8156b0eac1ea5820e780c0a5aa

K. Demeter, Z. Matyusz, The impact of lean practices on inventory turnover. Int. J. Prod. Econ. **133**(1), 154–163 (2011). https://doi.org/10.1016/j.ijpe.2009.10.031

J.R. Díaz-Reza, J.L. García-Alcaraz, J.R. Mendoza-Fong, V. Martínez-Loya, E.J. Macías, J. Blanco-Fernández, Interrelations among SMED stages: a causal model. Complexity (2017). https://doi.org/10.1155/2017/5912940

S.M. Disney, D.R. Towill, The effect of vendor managed inventory (VMI) dynamics on the bullwhip effect in supply chains. Int. J. Prod. Econ. **85**(2), 199–215 (2003). https://doi.org/10.1016/S0925-5273(03)00110-5

A. Formigoni, J.G.M. dos Reis, R.P. Moia, C.F. Stettiner, J.R. Maiellaro, Effectiveness of vendor managed inventory-VMI: a study applied in a mining company. Paper presented at the IFIP international conference on advances in production management systems, 2020

D.C. Freitas, R.N. Tomas, R.L.C. Alcantara, Estoque gerenciado pelo fornecedor (VMI): análise das barreiras e fatores críticos de sucesso em empresas de grande porte. Revista De Administração Unimep **11**(3), 221–253 (2013)

M. García, C. Quispe, L. Ráez, Mejora continua de la calidad en los procesos. Ind. Data **6**(1), 89–94 (2003)

J.A. Garza-Reyes, V. Kumar, S. Chaikittisilp, K.H. Tan, The effect of lean methods and tools on the environmental performance of manufacturing organisations. Int. J. Prod. Econ. **200**, 170–180 (2018). https://doi.org/10.1016/j.ijpe.2018.03.030

V. Gaur, M.L. Fisher, A. Raman, An econometric analysis of inventory turnover performance in retail services. Manage. Sci. **51**(2), 181–194 (2005). https://doi.org/10.1287/mnsc.1040.0298

S.B. Gershwin, *Manufacturing Systems Engineering* (Prentice-Hall, NJ, 1994)

J. Gosling, L. Purvis, M.M. Naim, Supply chain flexibility as a determinant of supplier selection. Int. J. Prod. Econ. **128**(1), 11–21 (2010). https://doi.org/10.1016/j.ijpe.2009.08.029

K. Govindan, D. Kannan, A.N. Haq, Analyzing supplier development criteria for an automobile industry. Ind. Manage. Data Syst. (2010)

A.D. GS, A. Indahingwati, N. Kurniasih, T. Listyorini, W. Fitriani, A.P. Utama, Planning and controlling of raw material inventory in efforts to avoid raw materials shortage. Int. J. Pure Appl. Mathe. **119**(17), 1977–1982 (2018)

L. Gualtieri, R.A. Rojas, M.A. Ruiz Garcia, E. Rauch, R. Vidoni, Implementation of a laboratory case study for intuitive collaboration between man and machine in sme assembly, in *Industry 4.0 for SMEs: Challenges, Opportunities and Requirements* (2020), pp. 335–382

D. Guzel, A.S. Asiabi, Improvement setup time by using SMED and 5S (An application in SME). Int. J. Sci. Technol. Res. **9**(1), 3727–3732 (2020). Retrieved from https://www.scopus.com/inw ard/record.uri?eid=2-s2.0-85078798697&partnerID=40&md5=a8669e3aa9d14396802c9375 88090209

A.M.V. Guzmán, K.M.M. González, M.F. Canales, M.T.V. Garza, F.B. de la Rosa, Design of a centralized warehouse layout and operation flow for the automotive industry: A simulation approach. Paper presented at the proceedings of the international conference on industrial engineering and operations management, 2018

R. Ham, 7—Seating layout and safety regulations, in *Theatres*. ed. by R. Ham (Architectural Press, 1987), pp. 45–49

F. Hamzeh, M. Al Hattab, L. Rizk, G. El Samad, S. Emdanat, Developing new metrics to evaluate the performance of capacity planning towards sustainable construction. J. Cleaner Prod. **225**, 868–882 (2019). https://doi.org/10.1016/j.jclepro.2019.04.021

Y. He, K. Keung Lai, H. Sun, Y. Chen, The impact of supplier integration on customer integration and new product performance: the mediating role of manufacturing flexibility under trust theory. Int. J. Prod. Econ. **147**(Part B), 260–270 (2014). https://doi.org/10.1016/j.ijpe.2013.04.044

Y. Higuchi, V.H. Nam, T. Sonobe, Sustained impacts of Kaizen training. J. Econ. Behav. Organ. **120**, 189–206 (2015). https://doi.org/10.1016/j.jebo.2015.10.009

W.J. Hopp, M.L. Spearman, *Factory physics* (Waveland Press, 2011)

M.S. Jamaliah, M.A. Mohd Hashim, A. Ismail, Improvements of worksite control for pull system. ARPN J. Eng. Appl. Sci. **11**(12), 7699–7705 (2016). Retrieved from https://www.scopus.com/ inward/record.uri?eid=2-s2.0-84977090541&partnerID=40&md5=949766f0d6a90a17728c1 6e5e1dc66ef

U.K. Jani, Good manufacturing practices (GMP):"planning for quality and control in microbiology", in *Frontier Discoveries and Innovations in Interdisciplinary Microbiology* (Springer, 2016), pp. 71–77

G. Jimenez, G. Santos, J.C. Sá, S. Ricardo, J. Pulido, A. Pizarro, H. Hernández, Improvement of productivity and quality in the value chain through lean manufacturing—a case study. Proc. Manuf. **41**, 882–889 (2019). https://doi.org/10.1016/j.promfg.2019.10.011

S. Kamaruddin, A.K. Zahid, A. Noor Siddiquee, Y.-S. Wong, The impact of variety of orders and different number of workers on production scheduling performance: a simulation approach. J. Manuf. Technol. Manag. **24**(8), 1123–1142 (2013). https://doi.org/10.1108/JMTM-12-2010-0083

A.-A. Karam, M. Liviu, V. Cristina, H. Radu, The contribution of lean manufacturing tools to changeover time decrease in the pharmaceutical industry. A SMED project. Proc. Manuf. **22**, 886–892 (2018). https://doi.org/10.1016/j.promfg.2018.03.125

C.M. Karuppan, Strategies to foster labor flexibility. Int. J. Product. Perform. Manag. **53**(6), 532–547 (2004). https://doi.org/10.1108/17410400410556192

D. Kazanskaia, Y. Shepilov, B. Madsen, *Adaptive production management for small-lot enterprise*, in Vol. 9266. Lecture Notes in Computer Science (including subseries Lecture Notes in Artificial Intelligence and Lecture Notes in Bioinformatics) (2015), pp. 157–168

B. Kennedy, Accelerating productivity: applying multitasking strategies to boost output and smooth production flow. Cutting Tool Eng. **55**(7), 30 (2003). Retrieved from https://www.scopus.com/ inward/record.uri?eid=2-s2.0-77949756937&partnerID=40&md5=9e1ee7a7d5c16f3279918aef 31df2da3

S.C. Kim, K.S. Shin, Negotiation model for optimal replenishment planning considering defects under the VMI and JIT environment. Asian J. Shipping Logistics **35**(3), 147–153 (2019)

O. Kimura, H. Terada, Design and analysis of pull system, a method of multi-stage production control. Int. J. Prod. Res. **19**(3), 241–253 (1981)

D.R. Kiran, Chapter 13—Total productive maintenance, in *Total Quality Management* (Butterworth-Heinemann, 2017), pp. 177–192

M. Kotabe, X. Martin, H. Domoto, Gaining from vertical partnerships: knowledge transfer, relationship duration, and supplier performance improvement in the US and Japanese automotive industries. Strateg. Manag. J. **24**(4), 293–316 (2003)

D.R. Krause, The antecedents of buying firms' efforts to improve suppliers. J. Oper. Manag. **17**(2), 205–224 (1999). https://doi.org/10.1016/S0272-6963(98)00038-2

H. Kumar Banga, R. Kumar, P. Kumar, A. Purohit, H. Kumar, K. Singh, Productivity improvement in manufacturing industry by lean tool. Mater. Today: Proc. **28**, 1788–1794 (2020). https://doi.org/10.1016/j.matpr.2020.05.195

R. Kumar, S.P. Singh, Cellular facility layout problem: a case of tower manufacturing industry. Manage. Environ. Qual.: Int. J. (2019)

N. Lakshminarayanan, K. Surendra Babu, S. Amin, A. Sulficker, M.N.A. Nassar, Design optimization of robotic work cell layout in automotive industries. Paper presented at the IOP conference series: materials science and engineering, 2020

Y. Li, S. Dong, Study on supplier selection of manufacturing in lean closed-loop supply chain. Paper presented at the proceedings of the ninth international conference on management science and engineering management, Berlin, Heidelberg, 2015

P. Lietaert, N. Billen, S. Burggraeve, Model-based multi-attribute collaborative production cell layout optimization. Paper presented at the proceedings of the 2019 20th international conference on research and education in mechatronics, REM 2019

Y.-H. Lin, J.-R. Shie, C.-H. Tsai, Using an artificial neural network prediction model to optimize work-in-process inventory level for wafer fabrication. Expert Syst. Appl. **36**(2, Part 2), 3421–3427 (2009). https://doi.org/10.1016/j.eswa.2008.02.009

J.-C. Lu, T. Yang, C.-Y. Wang, A lean pull system design analysed by value stream mapping and multiple criteria decision-making method under demand uncertainty. Int. J. Comput. Integr. Manuf. **24**(3), 211–228 (2011). https://doi.org/10.1080/0951192X.2010.551283

Y.-H. Ma, Y. Koren, Operation of manufacturing systems with work-in-process inventory and production control. CIRP Ann. **53**(1), 361–365 (2004). https://doi.org/10.1016/S0007-8506(07)60717-3

K. Mahmood, E. Shevtshenko, Productivity improvement by implementing lean production approach. Paper presented at the proceedings of the international conference of DAAAM baltic "industrial engineering", 2015

A. Memari, S.M. Zahraee, A. Anjomanshoae, A.R.B.A. Rahim, Performance assessment in a production-distribution network using simulation. Caspian J. Appl. Sci. Res. **2**(5)

N.D. Minh, N.D. Nguyen, P.K. Cuong, Applying lean tools and principles to reduce cost of waste management: an empirical research in Vietnam. Manage. Prod. Eng. Rev. **10**(1), 37–49 (2019). https://doi.org/10.24425/mper.2019.128242

R.P. Mishra, G. Gupta, A. Sharma, Development of a model for total productive maintenance barriers to enhance the life cycle of productive equipment. Proc. CIRP **98**, 241–246 (2021). https://doi.org/10.1016/j.procir.2021.01.037

Montgomery, D. C. (2020). *Introduction to statistical quality control*: John Wiley & Sons.

H. Müller, J. Sladojevic, Rapid tooling approaches for small lot production of sheet-metal parts. J. Mater. Process. Technol. **115**(1), 97–103 (2001)

A.E. Mulyana, I. Zuliana, Economic order quantity method approach in raw material inventory control for a small medium enterprise. Paper presented at the international conference on applied science and technology 2019-social sciences track (iCASTSS 2019), 2019

S. Nakajima, *Introduction to TPM: total productive maintenance* (Translation) (Productivity Press, Inc., 1988), 129

S. Nallusamy, Productivity enhancement in a small scale manufacturing unit through proposed line balancing and cellular layout. Int. J. Performability Eng. **12**(6), 523–534 (2016)

G. Nawanir, T. Lim Kong, N. Othman Siti, Lean manufacturing practices in Indonesian manufacturing firms: are there business performance effects? Int. J. Lean Six Sigma **7**(2), 149–170 (2016). https://doi.org/10.1108/IJLSS-06-2014-0013

G. Nawanir, T. Lim Kong, N. Othman Siti, A.Q. Adeleke, Developing and validating lean manufacturing constructs: an SEM approach. Benchmarking: Int. J. **25**(5), 1382–1405 (2018). https://doi.org/10.1108/BIJ-02-2017-0029

T.G. Noordewier, G. John, J.R. Nevin, Performance outcomes of purchasing arrangements in industrial buyer-vendor relationships. J. Mark. **54**(4), 80–93 (1990)

C. Pantoja, J.P. Orejuela, J.J. Bravo, Metodología de distribución de plantas en ambientes de agrupación celular. Estudios Gerenciales **33**(143), 132–140 (2017). https://doi.org/10.1016/j.estger.2017.03.003

L. Pattanaik, B. Sharma, Implementing lean manufacturing with cellular layout: a case study. Int. J. Adv. Manuf. Technol. **42**(7–8), 772–779 (2009)

S. Paul, A. Azeem, An artificial neural network model for optimization of finished goods inventory. Int. J. Ind. Eng. Comput. (2011)

G. Pinto, F.J.G. Silva, A. Baptista, N.O. Fernandes, R. Casais, C. Carvalho, TPM implementation and maintenance strategic plan—a case study. Proc. Manuf. **51**, 1423–1430 (2020). https://doi.org/10.1016/j.promfg.2020.10.198

J. Puche, J. Costas, B. Ponte, R. Pino, D. de la Fuente, The effect of supply chain noise on the financial performance of Kanban and Drum-Buffer-Rope: an agent-based perspective. Expert Syst. Appl. **120**, 87–102 (2019). https://doi.org/10.1016/j.eswa.2018.11.009

R.G. Qiu, Virtual production line based WIP control for semiconductor manufacturing systems. Int. J. Prod. Econ. **95**(2), 165–178 (2005)

D. Reynolds, Inventory-turnover analysis: Its importance for on-site food service. Cornell Hotel Restaur. Adm. Q. **40**(2), 54–58 (1999). https://doi.org/10.1016/S0010-8804(99)80025-1

M.H.M. Rusli, A. Jaffar, S. Muhamud-Kayat, M.T. Ali, Selection criterion of production methods used in the Kanban Pull System at Malaysian auto suppliers. Paper presented at the IEOM 2015—5th international conference on industrial engineering and operations management, proceeding, 2015

S.S. Saghiri, V. Mirzabeiki, Buyer-led environmental supplier development: can suppliers really help it? Int. J. Prod. Econ. **233**, 107969 (2021). https://doi.org/10.1016/j.ijpe.2020.107969

K. Salonitis, C. Tsinopoulos, Drivers and barriers of lean implementation in the Greek manufacturing sector. Proc. CIRP **57**, 189–194 (2016). https://doi.org/10.1016/j.procir.2016.11.033

E.F. Sánchez, L.A. Camarero, M.F. Barcala, *Estrategia de producción* (McGraw-Hill, 2006)

K. Sari, Exploring the benefits of vendor managed inventory. Int. J. Phys. Distrib. Logistics Manage. (2007)

S. Shingo, *Una revolucion en la produccion: el sistema SMED, 3a Edicion* (Routledge, 2017)

H.A. Simon, Theories of bounded rationality. Decision Organ. **1**(1), 161–176 (1972)

R. Singh, A.M. Gohil, D.B. Shah, S. Desai, Total productive maintenance (TPM) implementation in a machine shop: a case study. Proc. Eng. **51**, 592–599 (2013). https://doi.org/10.1016/j.proeng.2013.01.084

T.P. Singh, Role of manpower flexibility in lean manufacturing. Paper presented at the proceedings of GLOGIFT-2008, Stevens Institute of Technology, Hoboken, NJ, 2008

T.P. Singh, Role of manpower flexibility in lean manufacturing, in *The Flexible Enterprise* (2014), pp. 309–319

C.V. Sunil Kumar, S. Routroy, R.K. Mishra, Lean supplier management for better cost structures. Mater. Today: Proc. **5**(9, Part 3), 18941–18945 (2018). https://doi.org/10.1016/j.matpr.2018.06.244

C. Sürie, B. Reuter, Supply chain analysis, in *Supply Chain Management and Advanced Planning: Concepts, Models, Software, and Case Studies*, ed. by H. Stadtler, C. Kilger, H. Meyr (Springer, Berlin, Heidelberg, 2015), pp. 29–54

R. Susanto, Raw material inventory control analysis with economic order quantity method. IOP Conf. Ser.: Mater. Sci. Eng. **407**, 012070 (2018). https://doi.org/10.1088/1757-899x/407/1/012070

A.A. Taleizadeh, M. Noori-daryan, Pricing, manufacturing and inventory policies for raw material in a three-level supply chain. Int. J. Syst. Sci. **47**(4), 919–931 (2016). https://doi.org/10.1080/00207721.2014.909544

Y. Tliche, A. Taghipour, B. Canel-Depitre, An improved forecasting approach to reduce inventory levels in decentralized supply chains. Eur. J. Oper. Res. **287**(2), 511–527 (2020). https://doi.org/10.1016/j.ejor.2020.04.044

P. N. T. Tran, M. Gorton, F. Lemke, When supplier development initiatives fail: Identifying the causes of opportunism and unexpected outcomes. J. Bus. Res. **127**, 277–289 (2021). https://doi.org/10.1016/j.jbusres.2021.01.009

N.C. Tsourveloudis, Y.A. Phillis, Fuzzy assessment of machine flexibility. IEEE Trans. Eng. Manage. **45**(1), 78–87 (1998)

B. Ulutas, An application of SMED methodology. World Acad. Sci. Eng. Technol.s **79**, 101 (2011)

B. Venkat Jayanth, P. Prathap, P. Sivaraman, S. Yogesh, S. Madhu, Implementation of lean manufacturing in electronics industry. Mater. Today: Proc. (2020). https://doi.org/10.1016/j.matpr.2020.02.718

J. Wei, Impact of supplier's lead time on strategic assembler-supplier relationship in pricing sensitive market. Paper presented at the proceedings of the thirteenth international conference on management science and engineering management, Cham, 2020

C.L. Wen, H.M. Wee, S. Wu, Revisiting lean manufacturing process with vendor managed inventory system. Paper presented at the proceedings of the 5th international Asia conference on industrial engineering and management innovation (IEMI2014), Paris, 2015

J.M. Whipple, D. Russell, Building supply chain collaboration: a typology of collaborative approaches. Int. J. Logistics Manage. (2007)

R.E. White, J.N. Pearson, J.R. Wilson, JIT manufacturing: a survey of implementations in small and large US manufacturers. Manage. Sci. **45**(1), 1–15 (1999)

E. White Richard, N. Pearson John, JIT, system integration and customer service. Int. J. Phys. Distrib. Logist. Manag. **31**(5), 313–333 (2001). https://doi.org/10.1108/EUM0000000005515

G. Yadav, S. Luthra, D. Huisingh, S.K. Mangla, B.E. Narkhede, Y. Liu, Development of a lean manufacturing framework to enhance its adoption within manufacturing companies in developing economies. J. Clean. Prod. **245**, 118726 (2020). https://doi.org/10.1016/j.jclepro.2019.118726

W. Yan, Y. Huang, Y. Wang, J. He, Supplier selection for equipment manufacturing under the background of free trade zone. Paper presented at the LISS 2014, Berlin, Heidelberg, 2015

Z.F. Zhang, D. Janet, An entropy-based approach for measuring the information quantity of small lots production in a flow shop. Zidonghua Xuebao/acta Automatica Sinica **46**(10), 2221–2228 (2020). https://doi.org/10.16383/j.aas.c180479

Y.F. Zhao, R. Brown, T.R. Kramer, X. Xu, Dimensional metrology for manufacturing quality control, in *Information Modeling for Interoperable Dimensional Metrology* (Springer, 2011), pp. 275–307

Chapter 3
Methodology

Abstract This chapter describes the materials and methods used to develop this research and validate the structural equation models presented here. The chapter is divided into different sections, according to the stages developed therein, starting with questionnaire development, its application to managers in companies, the data capture, and validation. Additionally, this chapter reports the items' integration into latent variables and the model and their interpretation and industrial implications in the decision-making process.

Keywords SEM · Structural equation models · Methodology · Data debugging · Model interpretation

3.1 Introduction

In this research, several activities have been executed, illustrated in Fig. 3.1. It starts with a literature review related to the leading lean manufacturing practices, which allows designing a questionnaire applied to the manufacturing industry. The information is captured in specialized statistical software, cleaned of extreme and missing values, and then validated each of the latent variables that intervene in the four structural equation models presented as a central part of this book.

Afterward, these validated variables are integrated into the models, and efficiency indexes are obtained iteratively, and if they are met, they are interpreted. In addition, one of the significant contributions of this book is that it reports a sensitivity analysis for each of the models presented, which is based on conditional probabilities that allow managers to know possible scenarios and facilitates the decision-making process by identifying risk variables, as well as those that favor the benefits they seek to obtain. Each of the activities developed as a methodology is illustrated in Fig. 3.1. Then, each of these activities is described in the following paragraphs in a deep way.

J. R. Díaz-Reza et al., *Best Practices in Lean Manufacturing*,
SpringerBriefs in Applied Sciences and Technology,
https://doi.org/10.1007/978-3-030-97752-8_3

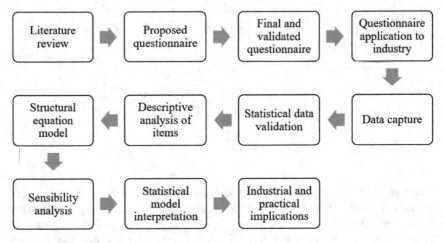

Fig. 3.1 Methodology

3.2 Literature Review

Since the objective of this book focuses on determining the relationship between manufacturing practices, a literature review was carried out to know the state of the art that existed concerning them. The review was performed in databases such as Springer, Ingenta, and Sciencedirect, among others, where the keywords are "manufacturing practices" and "lean manufacturing best practices". The articles identified were saved, and a database was created in which the name of the first author, the author's affiliation, the best manufacturing practices, and benefits identified, the country, among others, were saved, which allowed identifying the leading research groups in the area and generating Pareto diagrams associated with the frequency in which they were cited. In addition, the respective reference was downloaded in the Endnote X9® software for later use.

However, here is essential to mention that the paper reported by Nawanir et al. (2018) has a central role in our research because they report a structural equation model and a validated survey with several LM practices and their items. Items reported here were the primary basis of this research.

3.3 Proposed Questionnaire

In the literature review, the questionnaire applied by Nawanir et al. (2018) was identified in which 10 of the leading lean manufacturing practices are included, which are considered latent variables and were integrated by other observed variables. The manufacturing practices and the number of items are as follow: Flexible Resources (7 items), Cellular layouts (8 items), Pull/Kanban system (6 items), Small-lot production

(7 items), Quick setups (7 items), Uniform production level (7 items), Quality control (8 items), Total productive maintenance (TPM) (7 items), Supplier networks (7 items) and Inventory minimization (7 items), so it is decided to use it in our research and complement it with others that are used in the industrial environment, making the necessary adaptations (Gagnon et al. 2018) and add the manufacturing practices and benefits identified in the literature review.

3.4 Final and Validated Questionnaire

Nawanir et al. (2018) proposed the questionnaire adapted and modified by adding the practices and benefits identified in the literature review. However, that questionnaire is validated and adapted to the geographical context of Mexican manufacturing industries. Seven managers laboring in the maquiladora sector and four academics support the judges' validation in two rounds of review, evaluating adequacy, clarity, consistency, and relevance; a final questionnaire was defined.

The questionnaire is presented to all respondents in English to broaden the responses since some managers in companies established in Mexico are foreigners. The final questionnaire is divided into sections, and readers can consult the Nawanir et al. (2018) paper for an original version. The questionnaire is divided into sections as follow:

1. Introduction. Indicates the objective research and the use that will be made to information obtained, requesting permission for the assessments to be analyzed and published, respecting their anonymity.
2. Demographic data. Where aspects associated with the company's industrial sector, the respondent's current position, years of experience in their job position, gender, among others, are investigated (Antoine and Guillaume 1984). This section is answered with multiple-choice and open-ended questions.
3. LM practices. This section is further divided into subsections, where the LM practices and the items that comprise them are listed, which must be answered on a five-point Likert scale, where they indicate the following: 1—never, 2—rarely, 3—sometimes, 4—often and 5—always (Vonglao 2017).
4. Acknowledgment: In this section, we thank the participants and request an e-mail to send the analysis results if they are interested in a final report with all findings.

3.5 Questionnaire Application to Industry

The survey is applied to the manufacturing industry in Mexico, and the following considerations are considered:

- The survey is directed to personnel directly involved in the production lines and who apply LM tools as engineers and managers in manufacturing, production, and supply chain.
- The companies must be recognized by the IMMEX (Manufacturing, Maquiladora, and Export Services Industry), which guarantees that the manufacturing sector is the respondent and has a high flow of information and goods, with a complex supply chain.
- Preferably, respondents should have at least one year of experience to ensure that they know their work environment and the results of applying LM practices in production lines.

Because of the above, stratified sampling is followed, as it seeks to obtain information from a particular sector (Singh et al. 2016; Arnab 2017). The questionnaire is answered through scheduled, personalized, face-to-face interviews to answer doubts regarding understanding the items in the questionnaire. If an interview is canceled for any reason on three occasions, it is skipped due to the time consumption required to obtain it, and a thank-you statement is issued.

3.6   Data Capture and Debugging

A database is designed in SPSS v.25® statistical software to capture the information obtained from the questionnaires (IBM 2019). Each row represents a case or questionnaire, while each latent variable is presented in columns. SPSS software is chosen because of its simplicity, friendly interphase, and accessibility (IBM 2019; Šebjan and Tominc 2015).

For each latent variable and its items, an acronym is generated to help identify them quickly and easily, as well as a label that allows their complete identification. In the database, before analyzing the information already captured, the following activities are carried out to debug it:

- Identification of missing values. If the percentage of unanswered items is greater than 10%, this questionnaire is not included in the analysis (Crambes and Henchiri 2019). If the percentage is lower than 10%, the missing values are replaced by the median since the data are on an ordinal scale (Dray and Josse 2015).
- Identification of disengaged respondents. The standard deviation for every case is estimated, and given that the rating scale is from 1 to 5 if a case has a standard deviation lower than 0.5, it is also eliminated from the analysis (Gnagnarella et al. 2018).
- Outlier identification. All items are standardized, and if some items have values greater than 4 in absolute value, then it is considered an extreme value and is replaced by the median as a measure of central tenure (Hoffman 2019; Kaneko 2018).

3.7   Statistical Data Validation

With the database and its observed variables cleaned, we proceed to validate the latent variables to know if the items are well assigned, so the following indexes are used (Kock 2018):

- R^2 and Adjusted R^2. They measure parametric predictive validity on the latent variables and values greater than 0.02 are expected (Evermann and Tate 2016).
- Q^2 is used to measure non-parametric predictive validity and values similar to R^2 and greater than zero.
- Composite reliability and Cronbach's alpha measure construct validity and value greater than 0.7 are expected (Adamson and Prion 2013; Kile et al. 2014).
- The average variance extracted (AVE) measures latent variables' discriminant validity, and its values should be greater than 0.5 (Lee 2019).
- Full collinearity variance inflation factor. It is used to measure collinearity within the latent variables and should have values less than 5, although preferably less than 3.3 (Kock 2019b).

It is essential to mention that many of these indexes are obtained iteratively since frequently, the elimination of some item can improve the Cronbach's alpha index or reduce the collinearity between them, so not all the items in a latent variable may be integrated into the analysis, since they may have been eliminated to improve these indexes.

3.8   Descriptive Analysis of Items

With a database debugged, a descriptive analysis for the items is performed, where the following values are obtained:

- Median as a measure of central tendency, since the data come from ratings obtained through the questionnaire answered on a five-point Likert scale (Iacobucci et al. 2015). A low median value indicates that activity is not performed, or the benefit is not obtained, while a high value indicates that this activity is always performed or that the benefit is always obtained.
- Interquartile range (IR) to measure dispersion is the difference between the third and first quartile and is also used due to the nature of the data (Tominaga et al. 2018; Kang and Lee 2005). A low value in the RI indicates that there is consensus among respondents regarding the mean value of the item; however, a high value indicates a lot of dispersion and little consensus regarding the item's true value.

3.9 Structural Equation Model

The structural equation modeling technique is used to quantify the relationships between latent variables (LM practices), which are hypothesized. It is decided to use this technique because it allows to assess the relationships between latent variables that in turn are integrated by items (Avelar-Sosa et al. 2018), but at the same time allows these variables to have different roles, that is, they can play a role as independent and dependent variables in the same model (Nitzl 2016).

In this case, the latent variables are represented by ellipses and the relationships between them by arrows. The origin of the arrow represents the independent latent variable, while the destination represents the dependent variable. In contrast, the observed items or variables that integrate the latent variables are represented by rectangles (Aktepe et al. 2015).

The partial least squares (PLS) technique is used to evaluate the SEM, which is widely recommended when a normal distribution of the data cannot be assured, they come from ratings given on a Likert scale, or when there is a small sample (Martínez-Loya et al. 2018).

Before interpreting the model, the following model efficiency indices are analyzed:

- Average path coefficient (APC) with $P < 0.05$. It measures the overall significance of the relationships between variables.
- Average R-squared (ARS) and Average adjusted R-squared (AARS) with $P = 0.05$ to measure the model's predictive value.
- Average block VIF (AVIF) and Average full collinearity VIF (AFVIF) that is acceptable if ≤ 5 and are used to measure collinearity.
- Tenenhaus GoF (GoF) with values acceptable if ≥ 0.25, which indicates the fit of the data to the model.

If the above efficiency indices are met in the model under analysis, the model interpretation proceeds. Three types of effects are measured in the structural equation model described below and evaluated with a confidence level of 95%.

Here it is essential to mention that not all the items in the questionnaire are integrated into the latent variables that make up the model since some of them may be eliminated in the validation process. For example, it is possible to have seven items in the questionnaire but only five after validation (two were eliminated).

3.9.1 Direct Effects

They represent the hypotheses posed in the structural equation model and are indicated by an arrow linking two latent variables. It is represented by an index β representing the intensity of change between the two variables analyzed and expressed in standardized units, avoiding dimensionality problems in the analysis (Farooq et al. 2018).

A p-value is associated with each β value to measure the statistical significance and allow conclusions regarding the acceptance or rejection of the hypotheses posed (Manjot Singh and Anjali 2018). The null hypothesis tested is H_0: $\beta = 0$ versus the alternative hypothesis H_1: $\beta \neq 0$. All hypotheses are tested with a confidence level of 95%. If after the statistical test, it is demonstrated that $\beta = 0$ then the conclusion is that there is not a relationship between variables; however, if $\beta \neq 0$, then there is a relationship between variables analyzed.

A graph is used to illustrate the direct effects before and after the evaluation. The graph before evaluation indicates the proposed hypotheses, and the graph after evaluation indicates the values obtained for β and the p-value associated. In addition, an R^2 value is illustrated for each dependent variable, indicating the variance explained by the latent independent variables related to it.

Also, the effect sizes (ES) are reported, which are a decomposition of the R^2 value according to the contribution of each independent variable, which allows identifying which are the most important independent variables and have the best explanatory power in a dependent variable for each relationship or hypothesis proposed (Verdam et al. 2017).

3.9.2 Sum of Indirect Effects and Total Effects

Relationships between latent variables can occur through mediating variables and are then called indirect effects so that they can be of two segments or more (Boch et al. 2018; Egerer et al. 2018). Here, the sum of all indirect effects is reported, and a standardized β value is associated with every relationship between latent variables. That β value has associated a p-value and in this case, the hypothesis test for every β is similar to the direct effects with a confidence level of 95%.

These indirect effects are necessary to identify hidden relationships between latent variables or measure mediating variables' impact. Sometimes the direct effect is not supported; however, the indirect is statistically significant, indicating the importance of a mediator variable that enforces the relationship (Intakhan 2014; Aboelmaged 2018).

In the same way, the ES values are reported for indirect effects as a measure of the explanatory power of the independent latent variable on the dependent variable. Also, only the sum of ES is reported for every indirect relationship.

Finally, the sum of direct and indirect effects represents the total effects between variables (Schubring et al. 2016). Also, a standardized β value is reported with a p-value for its statistical significance test. In addition, ES is added as a decomposition of the value of the R^2 value in the latent dependent variable.

3.10 Sensitivity Analysis

Since the estimations by the partial least squares technique are based on standardized values, then in each of the evaluated models, a sensitivity analysis is reported for low (−) and high (+) scenarios of the latent variables in each of the relationships (Kock 2019a, 2018). Thus, three probabilities are analyzed:

- The probability that the independent and dependent latent variable appears at their low and high levels in isolation.
- The probability that the dependent and independent latent variables appear together in their high and low levels for their four combinations (+, +), (+, −), (−, +), (−, +), (−, −). This probability is represented by the symbol &.
- The conditional probability that the latent dependent variable will occur at a certain level (high or low), given that the independent variable has occurred at another (high or low). The conditional probabilities of the combinations (+, +), (+, −), (−, +), (−, −), (−, −) are analyzed and represented by If. This probability is helpful since it helps to identify risks and critical factors in the log of some benefits analyzed as response variables.

3.11 Statistical Model Interpretation

Each model is evaluated with a confidence level of 95%, which indicates that the significance level is 5%. A p-value is associated with each effect analyzed (direct, indirect, and total) to determine whether the relationship between the variables involved is significant or not, which allows the hypotheses to be accepted.

References

M. Aboelmaged, Direct and indirect effects of eco-innovation, environmental orientation and supplier collaboration on hotel performance: an empirical study. J. Clean. Prod. **184**, 537–549 (2018). https://doi.org/10.1016/j.jclepro.2018.02.192

K.A. Adamson, S. Prion, Reliability: measuring internal consistency using Cronbach's α. Clin. Simul. Nurs. **9**(5), e179–e180 (2013). https://doi.org/10.1016/j.ecns.2012.12.001

A. Aktepe, S. Ersöz, B. Toklu, Customer satisfaction and loyalty analysis with classification algorithms and structural equation modeling. Comput. Ind. Eng. **86**, 95–106 (2015). https://doi.org/10.1016/j.cie.2014.09.031

P. Antoine, A. Guillaume, Using the method of dynamic clusters in the course of a socio-demographic survey. Stateco **38**, 46–57 (1984)

R. Arnab, Chapter 7—Stratified sampling, in *Survey Sampling Theory and Applications,* ed. by R. Arnab (Academic Press, 2017), pp 213–256. https://doi.org/10.1016/B978-0-12-811848-1.000 07-8

L. Avelar-Sosa, J.L. Garcia-Alcaraz, A.A. Maldonado-Macias, J.M. Mejia-Munoz, Application of structural equation modelling to analyse the impacts of logistics services on risk perception, agility and customer service level. Adv. Prod. Eng. Manage. **13**(2), 179–192 (2018). https://doi.org/10.14743/apem2018.2.283

S. Boch, E. Allan, J.-Y. Humbert, Y. Kurtogullari, M. Lessard-Therrien, J. Müller, D. Prati, N.S. Rieder, R. Arlettaz, M. Fischer, Direct and indirect effects of land use on bryophytes in grasslands. Sci. Total Environ. **644**, 60–67 (2018). https://doi.org/10.1016/j.scitotenv.2018.06.323

C. Crambes, Y. Henchiri, Regression imputation in the functional linear model with missing values in the response. J. Stat. Plann. Inference **201**, 103–119 (2019). https://doi.org/10.1016/j.jspi.2018.12.004

S. Dray, J. Josse, Principal component analysis with missing values: a comparative survey of methods. Plant Ecol. **216**(5), 657–667 (2015). https://doi.org/10.1007/s11258-014-0406-z

M.H. Egerer, H. Liere, B.B. Lin, S. Jha, P. Bichier, S.M. Philpott, Herbivore regulation in urban agroecosystems: direct and indirect effects. Basic Appl. Ecol. **29**, 44–54 (2018). https://doi.org/10.1016/j.baae.2018.02.006

J. Evermann, M. Tate, Assessing the predictive performance of structural equation model estimators. J. Bus. Res. **69**(10), 4565–4582 (2016). https://doi.org/10.1016/j.jbusres.2016.03.050

M.S. Farooq, M. Salam, A. Fayolle, N. Jaafar, K. Ayupp, Impact of service quality on customer satisfaction in Malaysia airlines: A PLS-SEM approach. J. Air Transp. Manag. **67**, 169–180 (2018). https://doi.org/10.1016/j.jairtraman.2017.12.008

F. Gagnon, T. Aubry, J.B. Cousins, S.C. Goh, C. Elliott, Validation of the evaluation capacity in organizations questionnaire. Eval. Program Plann. **68**, 166–175 (2018). https://doi.org/10.1016/j.evalprogplan.2018.01.002

P. Gnagnarella, D. Dragà, A.M. Misotti, S. Sieri, L. Spaggiari, E. Cassano, F. Baldini, L. Soldati, P. Maisonneuve, Validation of a short questionnaire to record adherence to the Mediterranean diet: an Italian experience. Nutr. Metab. Cardiovasc. Dis. **28**(11), 1140–1147 (2018). https://doi.org/10.1016/j.numecd.2018.06.006

J.I.E. Hoffman, Chapter 9—Outliers and extreme values, in *Basic Biostatistics for Medical and Biomedical Practitioners*, ed. by J.I.E. Hoffman, 2nd edn (Academic Press, Boston, MA, USA, 2019), pp. 149–155. https://doi.org/10.1016/B978-0-12-817084-7.00009-7

D. Iacobucci, S.S. Posavac, F.R. Kardes, M.J. Schneider, D.L. Popovich, Toward a more nuanced understanding of the statistical properties of a median split. J. Consum. Psychol. **25**(4), 652–665 (2015). https://doi.org/10.1016/j.jcps.2014.12.002

IBM, IBM SPSS Statistics for Windows, 25.0 edn (IBM Corporation Armonk, NY, USA, 2019)

P. Intakhan, Direct & indirect effects of top management support on abc implementation success: evidence from ISO 9000 certified companies in Thailand. Proc. Soc. Behav. Sci. **164**, 458–470 (2014)

H. Kaneko, Automatic outlier sample detection based on regression analysis and repeated ensemble learning. Chemom. Intell. Lab. Syst. **177**, 74–82 (2018). https://doi.org/10.1016/j.chemolab.2018.04.015

S.J. Kang, M. Lee, Q-convergence with interquartile ranges. J. Econ. Dyn. Control **29**(10), 1785–1806 (2005). https://doi.org/10.1016/j.jedc.2004.10.004

H. Kile, K. Uhlen, G. Kjølle, Scenario selection in composite reliability assessment of deregulated power systems. Int. J. Electr. Power Energy Syst. **63**, 124–131 (2014). https://doi.org/10.1016/j.ijepes.2014.05.071

N. Kock, *WarpPLS 6.0 User Manual* (ScriptWarp Systems, Laredo, TX, USA, 2018)

N. Kock, Factor-based structural equation modeling with WarpPLS. Australasian Market. J. (AMJ) (2019). https://doi.org/10.1016/j.ausmj.2018.12.002

N. Kock, Factor-based structural equation modeling with WarpPLS. Australasian Market. J. (AMJ) **27**(1), 57–63 (2019). https://doi.org/10.1016/j.ausmj.2019.02.002

D. Lee, The convergent, discriminant, and nomological validity of the depression anxiety stress scales-21 (DASS-21). J. Affect. Disord. (2019). https://doi.org/10.1016/j.jad.2019.06.036

B. Manjot Singh, A. Anjali, Assessing relationship between quality management systems and business performance and its mediators : SEM approach. Int. J. Qual. Reliab. Manage. **8**, 1490 (2018). https://doi.org/10.1108/IJQRM-05-2017-0091

V. Martínez-Loya, J.R. Díaz-Reza, J.L. García-Alcaraz, J.Y. Tapia-Coronado, SEM: a global technique—Case applied to TPM, in *New Perspectives on Applied Industrial Tools and Techniques*,
eds. by J.L. García-Alcaraz, G. Alor-Hernández, A.A. Maldonado-Macías, C. Sánchez-Ramírez
(Springer International Publishing, Cham, 2018), pp. 3–22. https://doi.org/10.1007/978-3-319-
56871-3_1

G. Nawanir, K.T. Lim, S.N. Othman, A.Q. Adeleke, Developing and validating lean manufacturing
constructs: an SEM approach. Benchmarking: Int. J. **25**(5), 1382–1405 (2018). https://doi.org/
10.1108/BIJ-02-2017-0029

C. Nitzl, The use of partial least squares structural equation modelling (PLS-SEM) in management
accounting research: directions for future theory development. J. Account. Lit. **37**, 19–35 (2016).
https://doi.org/10.1016/j.acclit.2016.09.003

S. Schubring, I. Lorscheid, M. Meyer, C.M. Ringle, The PLS agent: predictive modeling with
PLS-SEM and agent-based simulation. J. Bus. Res. **69**(10), 4604–4612 (2016). https://doi.org/
10.1016/j.jbusres.2016.03.052

U. Šebjan, P. Tominc, Impact of support of teacher and compatibility with needs of study on
usefulness of SPSS by students. Comput. Hum. Behav. **53**, 354–365 (2015). https://doi.org/10.
1016/j.chb.2015.07.022

S. Singh, S.A. Sedory, M. del Mar Rueda, A. Arcos, R. Arnab, 8—Tuning in stratified sampling,
in *A New Concept for Tuning Design Weights in Survey Sampling*, eds. by S. Singh, S.A. Sedory,
M. del Mar Rueda, A. Arcos, R. Arnab (Academic Press, 2016), pp 219–256. https://doi.org/10.
1016/B978-0-08-100594-1.00008-5

R. Tominaga, M. Sekiguchi, K. Yonemoto, T. Kakuma, S.-i Konno, Establishment of reference
scores and interquartile ranges for the Japanese Orthopaedic Association Back Pain Evaluation
Questionnaire (JOABPEQ) in patients with low back pain. J. Orthop. Sci. **23**(4), 643–648 (2018).
https://doi.org/10.1016/j.jos.2018.03.010

M.G.E. Verdam, F.J. Oort, M.A.G. Sprangers, Structural equation modeling–based effect-size
indices were used to evaluate and interpret the impact of response shift effects. J. Clin. Epidemiol.
85, 37–44 (2017). https://doi.org/10.1016/j.jclinepi.2017.02.012

P. Vonglao, Application of fuzzy logic to improve the Likert scale to measure latent variables.
Kasetsart J. Soc. Sci. **38**(3), 337–344 (2017). https://doi.org/10.1016/j.kjss.2017.01.002

Chapter 4
Model 1. Distribution and Maintenance

Abstract This chapter presents a structural equation model integrated by four latent variables that are related using six hypotheses. The independent variables are *Cell layout* (*CLA*), *Total productive maintenance* (*TPM*), and *Single-minute exchange of die* (*SMED*), and the dependent variable is *Inventory minimization* (*INMI*). The model is evaluated using the partial least squares technique with information from 228 responses to a questionnaire applied to the manufacturing industry. The direct effects, the sum of indirect and total effects are analyzed. The results indicate that *TPM* and *SMED* have the most significant explanatory power on *INMI* since they allow the flow of materials throughout the production process.

Keywords SEM · Cellular layout · TPM · SMED · Inventory minimization

4.1 Model Variables and Validation

This model consists of four variables: As latent independent variables are *Cell layout* (*CLA*) with four items, *Total productive maintenance* (*TPM*) with six items, and *Single-minute exchange of die* (*SMED*) with six items, and the dependent variable is *Inventory minimization* (*INMI*) with five items, for a total of 21 items. The items analyzed are those resulting from the validation process, which is described below.

Table 4.1 illustrates the validation indexes for the four latent variables analyzed. The first row indicates the number of items before and after the validation process. Observe that some items were eliminated during the validation process, and that is why the initial and final number of items do not coincide.

According to *R*-squared and adjusted *R*-squared, it is concluded that there is sufficient parametric predictive validity since their values are greater than 0.02, while according to *Q*-squared, it is concluded that there is non-parametric predictive validity. Likewise, given the values of Cronbach's alpha and the composite reliability index, it is concluded that there is sufficient content validity.

Likewise, it is observed that there are no collinearity problems within the latent variables, given that the VIFs are less than 3.3 and that there is sufficient convergent

Table 4.1 Latent variable validation

	CLA		TPM		SMED		INMI	
Items	8	4	7	6	7	6	7	5
R-squared			0.192		0.353		0.353	
Adjusted R-squared	0.188		0.347		0.344			
Composite realiability	0.81		0.897		0.861		0.95	
Cronbach's alpha	0.688		0.861		0.805		0.934	
Average variance extracted	0.518		0.594		0.509		0.794	
Full collinearity VIF	1.394		1.572		1.659		1.457	
Q-squared			0.195		0.354		0.355	

validity, given that the average variance extracted is greater than 0.5 for all the variables.

For readers interested in other validation indexes, please review Annex at the end of the chapter.

4.2 Descriptive Analysis

Table 4.2 illustrates the descriptive analysis of the items remaining in the latent variables after the validation process. Each item has been ordered in descending order to facilitate the understanding of the univariate analysis. In general, it is observed that all items, except for one, have a median greater than four. This low median item is related to daily maintenance systems.

Thus, according to the *Cell layout* variable, it is observed that the most important thing is that the machines are arranged concerning each other to minimize the handling of materials since that is where most accidents and losses of raw material occur. Regarding *TPM*, the most important thing for the respondents is that a period is dedicated to carrying out machine maintenance activities during the workday.

Concerning *SMED*, according to the median, the most important thing is that all tools are allocated in a special place for them, which is the basis of another philosophy called 5S. Finally, the most critical thing about *Inventory minimization* is to use general-purpose machines that perform various activities after some adjustments, which allows a continuous flow of materials without being stored next to them.

Finally, concerning the interquartile range as a measure of dispersion, it is observed that three of the items have values less than two, and all of them in the *TPM* variable, which indicates that the respondents give high priority to equipment maintenance activities to avoid having inventory in the process.

Table 4.2 Descriptive analysis

Laten variable/item	Median	IR
Cell layout		
Are production facilities arranged in relation to each other, so that material handling is minimized?	4.55	2.13
Are production processes located close together so that material movement is minimized?	4.46	2.32
Is different equipment grouped together at a workstation to process a family of parts with similar requirements (such as shapes, processing, or routing requirements)?	4.39	2.24
Do product families determine the factory layout?	4.36	2.15
Total productive maintenance		
We dedicate periodic inspection to keep machines in operation	4.97	1.76
We ensure that machines are in a high state of readiness for production all the time	4.72	1.91
We scrupulously clean workspaces (including machines and equipment) to make unusual occurrences noticeable	4.68	2.11
We emphasize a good maintenance system as a strategy for achieving quality compliance	4.66	1.98
We have time reserved each day for maintenance activities	4.40	2.49
We have a sound system of daily maintenance to prevent machine breakdowns from occurring	3.95	2.47
Single-minute exchange of die		
Do we emphasize putting all tools in a standard storage location?	4.59	2.06
Production workers do not have trouble in finding the equipment they need	4.58	2.20
Our production are workers trained on machines' setup activities?	4.51	2.07
Do we can quickly perform our machines' setup if there is a change in process requirements?	4.46	2.23
We aggressively work on reducing machines' setup times?	4.18	2.78
Production workers perform their own machines' setups	4.07	3.03
Inventory minimization		
We use general-purpose machines, which can perform several essential functions?	4.33	2.04
If a particular workstation has no demand, can production workers go elsewhere in the manufacturing facility to operate a workstation that has demand?	4.27	2.00

(continued)

Table 4.2 (continued)

Laten variable/item	Median	IR
Are production workers cross-trained to perform several different jobs?	4.27	2.19
If one production worker is absent, another production worker can perform the same responsibilities?	4.22	2.13
Are production workers capable of performing several different jobs?	4.10	2.14

4.3 Hypotheses in the Model

The objective of this model is to determine to what extent the variables associated with *CLA*, *TPM*, and *SMED* generate *INMI*. It is assumed that *CLA* is on the upper right-hand side of the model and on which all the other latent variables depend due to the cost of making new distribution adjustments, while *INMI* is on the lower left-hand side and on which all the other variables affect. The variables are integrated into six hypotheses and are justified in the following paragraphs.

Many benefits can be obtained from *CLA* in a production system, one of them is that it facilitates *TPM*, since it concentrates machines and tools in a single place or small space, with the advantage that many of them have a similar operation, so that those responsible for maintenance specialize in keeping them in optimal conditions (Lago et al. 2019). In the same way, employees are fully aware of the equipment they operate, their needs and perform periodic inspections to identify possible failures (McWilliam et al. 2018), which makes it possible to anticipate production system failures and, many times, to prevent occupational accidents due to equipment malfunctions (Azadeh et al. 2013).

However, to achieve such human–machine interaction, extensive training is required in the use and operation of the machines and the personnel associated with the maintenance and service life of equipment parts and components (Vitayasak et al. 2019). Also, given a high level of synchronization of workstations in the *CLA*, special times should be dedicated simultaneously to the maintenance of machines, which allows maintaining the flow of materials in the production system continuously (McWilliam et al. 2018). Therefore, the following hypothesis is proposed in this research:

H_1. In a production system, *Cell layout* has a direct and positive effect on *Total productive maintenance*.

Cell layout has been associated with employee safety since employees are highly trained, and authors such as Akright and Kroll (1998) indicated more than 20 years ago that the number of accidents and rapid changeovers were an indication that the distribution of machines in the production system should be revised. In addition, Tayal et al. (2020) indicate that one of the advantages of the *Cell layout* is precisely the ease with which product changes can be made, since these changes are carried out in a single unit, without affecting many production lines.

Also, Vijay and Gomathi Prabha (2020) indicate that the grouping of machines in a *Cell layout* has made it possible to reduce waste, and one of the most important is that associated with rapid changes in production lines since it minimizes the time in which machines and operators remain without performing activities. In addition, as mentioned by Chu et al. (2019), operators in a *Cell layout* are more skilled and specialized, so tooling changes are made quickly. Finally, Morales Méndez and Rodriguez (2017) indicate that *SMED* has been widely favored in *Cell layout*. For the above, the following hypothesis is proposed:

H_2. *Cell layout* in a production system has a direct and positive impact on the *Single-minute exchange of die.*

Some authors, such as Díaz-Reza et al. (2017), consider *SMED* an essential part of *TPM* since they consider that the latter is responsible for minimizing setup times between one product or another. Other research, such as that reported by Bevilacqua et al. (2015), indicates that it is possible to reduce changeover times by applying many lean manufacturing tools, but one of the most important is *TPM*.

Also, a review performed by da Silva and Godinho Filho (2019) on *SMED* and the relationships it has with other LM tools indicates that *TPM* is the one with which it is best integrated and where it offers benefits associated with the safety of workers since, by their level of specialty, everyone knows how to do their activities, so the following hypothesis is proposed:

H_3. *Total productive maintenance* in a production system has a direct and positive impact on the execution of *Single-minute exchange of die.*

No doubt, making adjustments and model changes quickly brings benefits in the flow of materials when there is a *CLA* since there is no material in process waiting to be worked or at least it is concentrated in a single place; however, there is little space for prolonged storage with this type of distribution (Rodríguez-Méndez et al. 2015).

The structure and organization of machinery and equipment in a production system depend on the volume and demand required, so it directly relates to inventory levels and supply chain efficiency rates (Aalaei and Davoudpour 2017). For example, if machines are distributed within a short distance from each other, Wang et al. (2017) have found that inventories between workstations are reduced, while Chu et al. (2019) mentions that groups of operators become experts in their operations and adequately know the flow of material, which increases the flow of material, reducing in-process inventory. Finally, Böllhoff et al. (2016) indicate that human errors decrease considerably due to their level of integration.

In the same way, cellular distribution forces inventories between processes or machines to be low since spaces are reduced, and there cannot be high volumes of material waiting to be processed (Liu et al. 2018). The above forces production system planners to make adjustments to increase inventory turnover, so manufacturing cells require special attention. Therefore, the following hypothesis is proposed:

H_4. Cellular layout in a production system has a direct and positive effect on *Inventory minimization.*

Although the objective of *TPM* is to keep all the machines in an operative state, one of the benefits obtained from it is that there is an *Inventory minimization* because the raw material that arrives at the work stations is always processed, without having to wait due to breakdowns (Logeshwaran et al. 2021). Likewise, Hooda and Gupta (2019) state that *TPM* is a tool that can increase the excellence of companies if it is appropriately applied, and one of the main benefits is that it reduces inventories waiting to be processed.

For their part, Arai and Sekine (2017) consider *TPM* as one of the essential tools to achieve a lean factory and propose a series of worksheets and methods that allow better management of the machines and equipment that the company has. Finally, Burawat (2016) states that *TPM* is one of the LM tools that help to improve productivity in several indexes, such as inventory turnover and the level of defects generated by poorly calibrated machines. Due to the above, the following hypothesis is proposed:

H_5. *Total productive maintenance* in a production system has a direct and positive effect on *Inventory minimization.*

Although the relationship of *Single-minute exchange of die* with *Inventory minimization* seems logical, it should be explained. The relationship occurs when *SMED* reduces preparation times from one product to another because raw materials reduce their waiting time to be processed (da Silva and Godinho Filho 2019), reducing bottlenecks and maintaining a constant flow in the production system (Stadnicka 2015).

In addition, Díaz-Reza et al. (2017) also mention that *SMED* also contributes to smaller physical spaces required to store the raw material to be processed since waiting times are negligible. Rodríguez-Méndez et al. (2015) indicate that SMD favors the continuous flow of production systems and facilitates the achievement of metrics associated with JIT. Therefore, the following hypothesis is proposed:

H_6. *Single-minute exchange of die* in a production system has a direct and positive effect with *Inventory minimization.*

Figure 4.1 illustrates the relationships between the variables as hypotheses.

4.4   Evaluation of the Structural Equation Model

Since the latent variables have passed the validation process, they are integrated into the structural equation model, evaluated according to the methodology described. The efficiency indexes of the model are shown below:

- Average path coefficient (APC) $= 0.309$, $P < 0.001$
- Average R-squared (ARS) $= 0.300$, $P < 0.001$

Fig. 4.1 Proposed
model—distribution and
maintenance

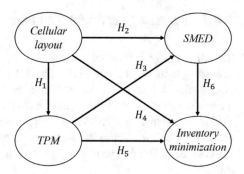

- Average adjusted R-squared (AARS) = 0.293, $P < 0.001$
- Average block VIF (AVIF) = 1.350, acceptable if ≤ 5, ideally ≤ 3.3
- Average full collinearity VIF (AFVIF) = 1.521, acceptable if ≤ 5, ideally ≤ 3.3
- Tenenhaus GoF (GoF) = 0.425, small ≥ 0.1, medium ≥ 0.25, large ≥ 0.36

According to the previous indexes, it is observed that since the APC index is associated with a p-value lower than the significance level, it is concluded that there is evidence to indicate that the β indexes in the model are generally adequate. At the same time, ARS and AARS are also associated with p-values lower than 0.05 and that there is sufficient predictive validity; likewise, the AVIF and AFVIF indices are less than 3.3, indicating that there are no collinearity problems between the variables analyzed, and finally, the GoF index is greater than 0.36, indicating that the data have a good fit to the model, and we proceed to its interpretation.

Figure 4.2 illustrates the model evaluated, where the β value is indicated for each of the relationships between variables, and p-values to determine the statistical significance of the model. Also, each latent dependent variable illustrates an R^2 value as a measure of the variance explained.

Fig. 4.2 Evaluated model

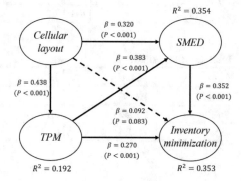

4.4.1   Direct Effect and Effect Size

The direct effects serve to validate the hypotheses established in Fig. 4.1, and according to the values indicated in Fig. 4.2, it is concluded that five relationships are statistically significant while one is not. Table 4.3 illustrates a summary regarding each of the effects. For example, the conclusion for H_1 is that there is sufficient statistical evidence to state with 95% confidence that *Cell layout* has a direct and positive effect on *Total productive maintenance* since when the first variable increases its standard deviation by one unit, the second increases its standard deviation by 0.438 units.

Figure 4.2 of the model evaluated shows that the latent dependent variables have an R^2 value to measure the variance explained by the independent latent variables. However, there are dependent variables that are explained by more than one dependent variable, so Table 4.4 presents a summary of the contributions of each of them. From this information, it is concluded that *SMED* is explained by 0.354, but *SMED* contributes with 0.155 and *TPM* with 0.199, so based on the effect size, it is concluded that *TPM* is the variable that most favors *SMED*. However, *INMI* is explained by 0.353, but *CLA* contributes with 0.034, *TPM* with 0.133, and *SMED* with 188, which allows us to conclude that *SMED* is the variable that most favors *INMI* and managers should focus their efforts on it.

Table 4.3   Summary of direct effects

Hi	Independent variable	Dependent variable	β	p-value	Conclusion
H_1	CLA	TPM	0.438	<0.001	Supported
H_2	CLA	SMED	0.32	<0.001	Supported
H_3	TPM	SMED	0.383	<0.001	Supported
H_4	CLA	INMI	0.092	0.083	No supported
H_5	TPM	INMI	0.27	<0.001	Supported
H_6	SMED	INMI	0.352	<0.001	Supported

Table 4.4   Effect size for R^2

	CLA	TPM	SMED	R^2
TPM	0.192			0.192
SMED	0.155	0.199		0.354
INMI	0.034	0.131	0.188	0.353

Table 4.5 Sum of indirect effects

	CLA	TPM
SMED	0.168 $(p < 0.001)$ ES $= 0.081$	
INMI	0.29 $(p < 0.001)$ ES $= 0.108$	0.135 $(p = 0.002)$ ES $= 0.065$

4.4.2 Sum of Indirect Effects

As shown in Fig. 4.2, the relationship between *CLA* and *INMI* is not statistically significant, but the effect may be indirect. Table 4.5 illustrates the sum of indirect effects between the variables, where the β value, the associated *p*-value, and the effect size (ES) are indicated.

It is observed in Table 4.5 that the sum of indirect effects between *CLA* and *INMI* is statistically significant; it is the largest of all with a value of 0.29 and occurs through *TPM* and *SMED* as mediating variables and can explain up to 0.108 of their variance.

4.4.3 Total Effects

The sum of the direct and indirect effects generates the total effects of the model, which are shown in Table 4.6. It is observed that all of them are statistically significant and that the largest of them is between *CLA* and *SMED*, which allows us to conclude that *CLA* favors the implementation and execution of *SMED*, which may be because all the machines are grouped in the same place, which leads to the acquisition of a certain level of skills to carry out the rapid changes.

Table 4.6 Total effects

	CLA	TPM	SMED
TPM	$\beta = 0.438$ $(p < 0.001)$ ES $= 0.192$		
SMED	$\beta = 0.488$ $(p < 0.001)$ ES $= 0.236$	$\beta = 0.383$ $(p < 0.001)$ ES $= 0.199$	
INMI	$\beta = 0.382$ $(p < 0.001)$ ES $= 0.142$	$\beta = 0.405$ $(p < 0.001)$ 0.197	$\beta = 0.352$ $(p < 0.001)$ 0.188

Table 4.7 Sensitivity analysis

			CLA		TPM		SMED	
			+	−	+	−	+	−
			0.150	0.177	0.173	0.177	0.168	0.168
TPM	+	0.173	& = 0.055 If = 0.364	& = 0.018 If = 0.103				
	−	0.177	& = 0.014 If = 0.091	& = 0.068 If = 0.385				
SMED	+	0.168	& = 0.073 If = 0.485	& = 0.005 If = 0.026	& = 0.077 If = 0.447	& = 0.000 If = 0.000		
	−	0.168	& = 0.009 If = 0.061	& = 0.064 If = 0.359	& = 0.027 If = 0.158	& = 0.064 If = 0.359		
INMI	+	0.123	& = 0.055 If = 0.364	& = 0.005 If = 0.026	& = 0.055 If = 0.316	& = 0.000 If = 0.000	& = 0.073 If = 0.432	& = 0.005 If = 0.027
	−	0.164	& = 0.000 If = 0.000	& = 0.050 If = 0.282	& = 0.005 If = 0.026	& = 0.059 If = 0.333	& = 0.000 If = 0.000	& = 0.036 If = 0.216

4.4.4 Sensitivity Analysis

Ideally, lean manufacturing tools should be at a high level of implementation. However, they can occur at low levels, and then the benefits of their implementation cannot be guaranteed, and since they are interrelated, they can affect each other. The sensitivity analysis allows different scenarios to be analyzed, where a "+" sign represents a high level while a "−" sign represents a low level. A low scenario is considered when the standardized variable has values less than −1, while a high scenario occurs when the standardized variable has values greater than 1.

Table 4.7 illustrates the sensitivity analysis for the model. Three types of probabilities are presented: that the variable occurs in isolation at its low and high levels, that two variables co-occur in a combination of scenarios, and that the dependent variable occurs given that the independent variable has occurred.

4.5 Conclusions and Industrial Implications

4.5.1 From the Structural Equation Model

From Fig. 4.2 and Tables 4.3, 4.4, and 4.5, regarding the direct, indirect, and total effects, the following can be concluded:

H_1. There is enough statistical evidence to state that, in a productive system, *Cell layout* has a direct and positive effect on *Total productive maintenance*, since when the first variable increases its standard deviation by one unit, the second increases it by 0.438.

H_2. There is sufficient statistical evidence to state that, *Cell layout* in a production system has a direct and positive impact on the execution of *Single-minute exchange of die* since when the first variable increases its standard deviation by one unit, the second one goes up by 0.320 units. In addition, an indirect effect is given through *Total productive maintenance* with a value of 0.168, which gives a total effect of 0.488.

H_3. There is sufficient statistical evidence to state that *Total productive maintenance* in a production system directly impacts the execution of *Single-minute exchange of die*. When the first variable increases its standard deviation by one unit, the second increases it by 0.383.

H_4. There is not enough statistical evidence to state that the cellular layout in a production system directly and positively affects *Inventory minimization* since the p-value associated with the β is 0.083, a value higher than the 0.05 established cut-off value. However, an indirect effect of 0.290 occurs through *Total productive maintenance* and *Single-minute exchange of die*, which is statistically significant, as is the total effect of 0.382.

H_5. There is sufficient statistical evidence to state that, *Total productive maintenance* in a production system has a direct and positive effect with *Inventory minimization* since when the first variable increases its standard deviation by one unit, the second one increases it by 0.270. In addition, there is an indirect effect through *Single-minute exchange of die* with a value of 0.135, which gives a total effect of 0.405.

H_6. There is sufficient statistical evidence to state that *Single-minute exchange of die* in a production system directly affects *Inventory minimization*. When the first variable increases its standard deviation by one unit, the second increases it by 0.352.

4.5.2 Conclusions of the Sensitivity Analysis

For the relationship $CLA \rightarrow TPM$, it is observed that $CLA+$ facilitates obtaining $TPM+$ with a probability of 0.364; furthermore, $CLA+$ is weakly associated with $TPM-$, since the probability is only 0.091, a low value. This indicates that managers should try to obtain high levels of CLA to guarantee the implementation and benefits of TPM.

However, that relationship shows that $CLA-$ can lead to $TPM-$ with a probability of 0.385, representing a risk for managers and decision-makers. The above is ratified when observing that $CLA-$ is associated very little with $TPM+$, with a probability of only 0.103, which indicates that, even if the favorable conditions of CLA are not present, there may be a slight chance of obtaining $TPM+$ because maintenance programs must exist regardless of the distribution that is present.

For the relationship of $CLA \to SMED$, it is observed that $CLA+$ is a significant facilitator of $SMED+$, as the conditional probability is 0.485; furthermore, $CLA+$ is only weakly associated with low levels of $SMED-$, as the conditional probability is only 0.061. The above indicates that managers can be confident that investments made in CLA will be favored by making rapid changes when switching from one product to another.

Likewise, in the above relationship, it is observed that $CLA-$ is not associated with $SMED+$, since the probability is only 0.005; however, scenarios in which $CLA-$ is present may represent a risk for managers since there is a 0.59 probability of generating $SMED-$. Thus, it is concluded that managers should seek to obtain $CLA+$ to avoid losing the benefits offered by $SMED+$.

Annex: Additional Data for Support Validation

T ratios for path coefficients

	CLA	TPM	SMED
TPM	7.046		
SMED	5.035	6.091	
INMI	1.389	4.213	5.57

Confidence intervals for path coefficients

	CLA		TPM		SMED	
	LCL	UCL	LCL	UCL	LCL	UCL
TPM	0.316	0.56				
SMED	0.196	0.445	0.26	0.506		
INMI	−0.038	0.222	0.145	0.396	0.228	0.476

T ratios for loadings

CL04	12.936			
CL07	13.180			
CL08	11.268			
CL06	11.235			
TPM01		15.056		
TPM02		14.449		
TPM03		11.707		
TPM04		12.315		

(continued)

(continued)

CL04	12.936			
TPM05		11.598		
TPM07		13.637		
CR04			10.662	
CR05			13.582	
CR06			13.204	
CR07			12.666	
CR02			11.103	
CR03			10.931	
MI01				15.606
MI02				16.026
MI03				15.947
MI04				16.445
MI05				13.672

Confidence intervals for loadings

	CLA		TPM		SMED		INMI	
	LCL	UCL	LCL	UCL	LCL	UCL	LCL	UCL
CL04	0.644	0.874						
CL07	0.657	0.886						
CL08	0.555	0.789						
CL06	0.553	0.787						
TPM01			0.753	0.979				
TPM02			0.722	0.949				
TPM03			0.579	0.811				
TPM04			0.611	0.842				
TPM05			0.573	0.806				
TPM07			0.681	0.909				
CR04					0.522	0.757		
CR05					0.678	0.906		
CR06					0.658	0.887		
CR07					0.630	0.860		
CR02					0.546	0.780		
CR03					0.537	0.771		
MI01							0.781	1.005
MI02							0.802	1.026
MI03							0.798	1.022

(continued)

(continued)

	CLA		TPM		SMED		INMI	
	LCL	UCL	LCL	UCL	LCL	UCL	LCL	UCL
MI04							0.823	1.046
MI05							0.682	0.911

PLSc reliabilities (Dijkstra's rho_a's)

CLA	TPM	SMED	INMI
0.751	0.873	0.822	0.937

Additional indices (indicator corr. matrix fit)

- Standardized root mean squared residual (SRMR) = 0.123, acceptable if ≤ 0.1
- Standardized mean absolute residual (SMAR) = 0.101, acceptable if ≤ 0.1
- Standardized chi-squared with 209 degrees of freedom (SChS) = 11.476, $P < 0.001$
- Standardized threshold difference count ratio (STDCR) = 0.905, acceptable if ≥ 0.7, ideally = 1
- Standardized threshold difference sum ratio (STDSR) = 0.769, acceptable if ≥ 0.7, ideally = 1

Additional reliability coefficients

	CLA	TPM	SMED	INMI
Dijkstra's PLSc reliability	0.751	0.873	0.822	0.937
True composite reliability	0.810	0.897	0.861	0.950
Factor reliability	0.810	0.897	0.861	0.950

Correlations among latent variables with squared roots of AVEs

	CLA	TPM	SMED	INMI
CLA	0.720	0.431	0.480	0.358
TPM	0.431	0.771	0.517	0.480
SMED	0.480	0.517	0.714	0.486
INMI	0.358	0.480	0.486	0.891

Full collinearity VIFs

CLA	TPM	SMED	INMI
1.394	1.572	1.659	**1.457**

HTMT ratios (good if < 0.90, best if < 0.85)

	CLA	TPM	SMED
TPM	0.551 ($p < 0.001$)		
SMED	0.641 ($p < 0.001$)	0.611	($p < 0.001$)
INMI	0.431 ($p < 0.001$)	0.543	($p < 0.001$)0.575 ($p < 0.001$)

References

A. Aalaei, H. Davoudpour, A robust optimization model for cellular manufacturing system into supply chain management. Int. J. Prod. Econ. **183**, 667–679 (2017). https://doi.org/10.1016/j.ijpe.2016.01.014

W.T. Akright, D.E. Kroll, Cell formation performance measures—determining when to change an existing layout. Comput. Ind. Eng. **34**(1), 159–171 (1998). https://doi.org/10.1016/S0360-8352(97)00158-7

K. Arai, K. Sekine, TPM for the lean factory: innovative methods and worksheets for equipment management (Taylor and Francis, Oxfordshire, United Kingdom, 2017). https://doi.org/10.1201/9780203735336

A. Azadeh, S. Motevali Haghighi, S.M. Asadzadeh, H. Saedi, A new approach for layout optimization in maintenance workshops with safety factors: the case of a gas transmission unit. J. Loss Prev. Process Ind. **26**(6), 1457–1465 (2013). https://doi.org/10.1016/j.jlp.2013.09.014

M. Bevilacqua, F.E. Ciarapica, I. De Sanctis, G. Mazzuto, C. Paciarotti, A changeover time reduction through an integration of lean practices: a case study from pharmaceutical sector. Assem. Autom. **35**(1), 22–34 (2015). https://doi.org/10.1108/AA-05-2014-035

J. Böllhoff, J. Metternich, N. Frick, M. Kruczek, Evaluation of the human error probability in cellular manufacturing. Proc. CIRP **55**, 218–223 (2016). https://doi.org/10.1016/j.procir.2016.07.080

P. Burawat, Guidelines for improving productivity, inventory, turnover rate, and level of defects in manufacturing industry. Int. J. Econ. Perspect. **10**(4), 88–95 (2016)

X. Chu, D. Gao, S. Cheng, L. Wu, J. Chen, Y. Shi, Q. Qin, Worker assignment with learning-forgetting effect in cellular manufacturing system using adaptive memetic differential search algorithm. Comput. Ind. Eng. **136**, 381–396 (2019). https://doi.org/10.1016/j.cie.2019.07.028

I.B. da Silva, M. Godinho Filho, Single-minute exchange of die (SMED): a state-of-the-art literature review. Int. J. Adv. Manuf. Technol. **102**(9), 4289–4307 (2019). https://doi.org/10.1007/s00170-019-03484-w

J.R. Díaz-Reza, J.L. García-Alcaraz, J.R. Mendoza-Fong, V. Martínez-Loya, E.J. Macías, J. Blanco-Fernández, Interrelations among SMED stages: a causal model. Complexity 2017. https://doi.org/10.1155/2017/5912940

A. Hooda, P. Gupta, Manufacturing excellence through total productive maintenance implementation in an indian industry: a case study. Int. J. Mech. Prod. Eng. Res. Dev. **9** 3), 1593–1604. https://doi.org/10.24247/ijmperdjun2019168

A. Lago, D. Trabucco, A. Wood, Chapter 7—Testing, inspection, and maintenance, in Damping Technologies for Tall Buildings eds. A. Lago, D. Trabucco, A. Wood (Butterworth-Heinemann, 2019), pp 465–531. https://doi.org/10.1016/B978-0-12-815963-7.00007-5

C. Liu, J. Wang, J.Y.T. Leung, Integrated bacteria foraging algorithm for cellular manufacturing in supply chain considering facility transfer and production planning. Appl. Soft Comput. **62**, 602–618 (2018). https://doi.org/10.1016/j.asoc.2017.10.034

J. Logeshwaran, R.M. Nachiappan, S. Nallusamy, Evaluation of overall manufacturing line effectiveness with inventory between sustainable processes in continuous product line manufacturing system. J. Green Eng. **11**(1), 104–121 (2021). https://doi.org/10.1108/17410380610688278

R. McWilliam, S. Khan, M. Farnsworth, C. Bell, Zero-maintenance of electronic systems: perspectives, challenges, and opportunities. Microelectron. Reliab. **85**, 122–139 (2018). https://doi.org/10.1016/j.microrel.2018.04.001

J.D. Morales Méndez, R.S. Rodriguez, Total productive maintenance (TPM) as a tool for improving productivity: a case study of application in the bottleneck of an auto-parts machining line. Int. J. Adv. Manuf. Technol. (2017). https://doi.org/10.1007/s00170-017-0052-4

R. Rodríguez-Méndez, D. Sánchez-Partida, J.L. Martínez-Flores, E. Arvizu-Barrón, A case study: SMED & JIT methodologies to develop continuous flow of stamped parts into AC disconnect assembly line in Schneider Electric Tlaxcala Plant. IFAC PapersOnLine **48**, 1399–1404 (2015). https://doi.org/10.1016/j.ifacol.2015.06.282

D. Stadnicka, Setup analysis: combining SMED with other tools. Manage. Prod. Eng. Rev. (MPER) **6**(1), 36 (2015). https://doi.org/10.1515/mper-2015-0006

A. Tayal, U. Kose, A. Solanki, A. Nayyar, J.A.M. Saucedo, Efficiency analysis for stochastic dynamic facility layout problem using meta-heuristic, data envelopment analysis and machine learning. Comput. Intell. **36**(1), 172–202 (2020). https://doi.org/10.1111/coin.12251

S. Vijay, M. Gomathi Prabha, Work standardization and line balancing in a windmill gearbox manufacturing cell: a case study. Material. Today: Proc. https://doi.org/10.1016/j.matpr.2020.08.584

S. Vitayasak, P. Pongcharoen, C. Hicks, Robust machine layout design under dynamic environment:dynamic customer demand and machine maintenance. Expert Syst. Appl. X:100015. https://doi.org/10.1016/j.eswax.2019.100015

P.-S. Wang, T. Yang, M.-C. Chang, Effective layout designs for the Shojinka control problem for a TFT-LCD module assembly line. J. Manuf. Syst. **44**, 255–269 (2017). https://doi.org/10.1016/j.jmsy.2017.07.004

Chapter 5
Model 2. Pull System and Quality Control

Abstract Product quality is one of the concerns of manufacturing managers, and the question they frequently ask themselves is, what should they do to guarantee it and achieve customer satisfaction? This paper presents a structural equation model (SEM), where Quality control (QUC) is the dependent variable, and it is assumed that Pull system (PUS), Small-lot production (SLP), and Uniform production level (UPL) can explain it. The variables are related using six hypotheses validated using the partial least squares technique in WarpPLS v.7.0 software, where a sensitivity analysis based on conditional probabilities is also reported. The results indicate that PUS is a lean manufacturing practice that has a high relationship with SLP. At the same time, SLP is associated with UPL, and finally, the latter is associated with QUC.

Keywords SEM · Pull system · Small-lot production · Uniform production level · Quality control

5.1 Model Variables and Their Validation

This model includes four variables, three of them independent and one dependent. The independent variables are *Pull system* (*PUS*) with four items, *Small-lot production* (*SLP*) with six items, and *Uniform production level* (*UPL*) with five items. The dependent variable is *Quality control* (*QUC*) with six items, integrating 21 items in the model. It is essential to mention that the number of items is the one that has resulted after the validation procedure, where some of them are eliminated to improve the indexes. Readers can consult the complete survey in the appendix of this book.

Table 5.1 describes the validation indices of the latent variables analyzed in this model. The first row shows that each latent variable has two numbers associated with the items. The first number refers to the number of items in the questionnaire, and the second is the remaining after the debugging process. In all the latent variables, at least one item has been eliminated to improve the indices.

Table 5.1 shows that the latent variables meet the cut-off values for each of the indices established in the methodology. For example, it is observed that all the variables have sufficient parametric predictive validity since the R^2 and adjusted R^2

© The Author(s), under exclusive license to Springer Nature Switzerland AG 2022
J. R. Díaz-Reza et al., *Best Practices in Lean Manufacturing*,
SpringerBriefs in Applied Sciences and Technology,
https://doi.org/10.1007/978-3-030-97752-8_5

Table 5.1 Latent variable validation

	PUS		SLP		UPL		QUC	
Items	6	4	7	6	7	5	7	6
R-squared			0.172		0.134		0.462	
Adjusted R-squared			0.168		0.126		0.454	
Composite realiability	0.892		0.886		0.866		0.885	
Cronbach's alpha	0.837		0.845		0.805		0.843	
Average variance extracted	0.673		0.566		0.566		0.562	
Full collinearity VIF	1.412		1.194		1.642		1.829	
Q-squared			0.195		0.354		0.464	

indices are greater than 0.02. At the same time, Q^2 indicates that there is sufficient parametric predictive validity. Likewise, the composite reliability and Cronbach's alpha indices are greater than 0.7, so it is concluded that there is sufficient internal validity in the latent variables.

Likewise, it is observed that the Average variance extracted is greater than 0.5 in all analyzed latent variables, which indicates that there is sufficient convergent validity. Finally, it is observed that the full collinearity VIF is lower than 3.3, which allows concluding that there are no collinearity problems within the variables analyzed.

Therefore, considering that the variables meet the validation indexes, they are integrated into the structural equation model for analysis.

Although the indices illustrated in Table 5.1 are the most important ones, Annex of this chapter illustrates other indices reported by the WarpPLS software that has been used so that users can verify other associated indices.

5.2 Descriptive Analysis of Items

Table 5.2 illustrates the descriptive analysis of the items for the latent variables of the model and indicates only those that remain after the validation process, and therefore may differ from the number contained in the questionnaire. The first column indicates the item name, the second column indicates the median value, and the third column indicates the interquartile range. For each latent variable, the items are ordered from highest to lowest according to the median values.

An overall analysis of the median values shows that only two items in the *Quality control* latent variable have values greater than or equal to 5, indicating that managers attach great importance to meeting customer specifications. Interestingly, the lowest median values are found in the *Small-lot production* variable, with a value of 3.31 and associated with the efforts made to reduce the lot size to be produced.

The most critical thing concerning *Pull system* items is that Kanban systems are used to authorize the production and material supply orders, making sense since LM

Table 5.2 Descriptive analysis for items

	Median	Interquartile range
Pull system		
Is the Kanban system used to authorize production? (Kanban is a work signaling system such as cards, verbal signals, light flashing, electronic messages, empty containers)	4.66	2.70
Is the Kanban system used to authorize material movements?	4.51	2.29
To authorize orders to suppliers, do we use supplier Kanban that rotates between factory and suppliers?	4.24	2.33
A pull system is used (producing in response to demand from the next stage of the production process) to control production rather than schedule prepared in advance?	4.22	2.33
Small-lot production		
Do we produce in a more frequent but smaller lot size?	4.01	2.45
Do we receive products from suppliers in a small-lot with frequent deliveries?	3.86	2.45
Do we emphasize producing a small number of items together in a batch?	3.70	2.33
In our production system, do we strictly avoid the flow of one item in large quantities together?	3.60	2.75
Do we emphasize producing in small-lot sizes to increase manufacturing flexibility?	3.58	2.27
Do we aggressively work on reducing production lot sizes?	3.31	2.37
Uniform production lot		
We emphasize equating workloads in each production process	4.70	1.74
Do we emphasize a more accurate forecast to reduce variability in production?	4.67	1.91
Each product is produced in a relatively fixed quantity per production period	4.57	1.77
Daily production of different product models is arranged in the same ratio with monthly demand	4.37	2.23
Do we produce by repeating the same combination of products from one day to another?	4.11	2.38
Quality control		
We use visual control systems (such as andon/line-stop alarm light, level indicator, warning signal, signboard) as a mechanism to make problems visible	5.16	1.84
Production processes on production floors are monitored with statistical quality control techniques	5.00	1.73
We have quality-focused teams that meet regularly to discuss quality issues	4.89	2.03

(continued)

Table 5.2 (continued)

	Median	Interquartile range
We use statistical techniques to reduce process variances	4.78	2.04
Quality problems can be traced to their source easily	4.57	2.09
Production workers are trained for quality control	4.51	2.03

seeks to reduce waste. The most critical item regarding *Small-lot production* is that it seeks to produce more frequently and smaller.

In a Uniform production lot, the most critical item is that the production lines are balanced to have a continuous flow, and emphasis is placed on reducing the variability of sales forecasts since this avoids sudden changes in programming. Finally, about *Quality control*, it is noted that the most important thing is to rely on visual aid systems that facilitate communicating when and where there is a problem in the production process, in addition to all production lines being monitored with statistical techniques.

5.3   Hypotheses in the Model

This model seeks to prove that the *Pull system* (*PUS*), *Small-lot production* (*SLP*), and *Uniform production level* (*UPL*) can influence the Quality Control (*QUC*) of a production process. The model integrates a total of six hypotheses, which are justified below.

In *PUS*, the entire flow of materials in the production system is demand-driven, and no product is manufactured if it has not been ordered. *PUS* is characterized by the fact that the warehouses or outlets are the ones that collect the daily sales information and generate the replenishment of stocks, which allows them to operate autonomously and, above all, with greater knowledge of cause (Kabadurmus and Durmusoglu 2019). Nevertheless, suppose there is adequate communication between outlets and manufacturers. In that case, it does not require a significant investment to forecast demand, which decreases the risk of demand variability (Hoga 2021).

However, this approach only considers the points of sale without knowing the needs and requirements of the production system, and it is likely that it does not have the required capacity to respond and may incur delivery time failures due to lack of communication within the company (Gillam et al. 2018). In addition, this type of production requires knowing the customer's demands in the short, medium, and long term, often unavailable. Sometimes, forecasts are risks due to possible accidents, political changes, natural disasters, and others (Guan et al. 2015).

To know if *PUS* is used in industry, in this research is indagated if Kanban system is used to: authorize the production, have a production according to demand, begin production only when users request a product, generate purchase orders, facilitate materials flow, among others (Takeda Berger et al. 2019; Ebner et al. 2019; Zhang et al. 2018).

Knowing the forecasts allows better planning and fulfillment of production orders on time and in small quantities. Thus, the *PUS* system in a production system can comply with the LMT called *Small-lot production*. Authors such as Fowler et al. (2019) indicate that *PUS* generates greater flexibility with small and time-varying production orders. However, Nelson (2016) states that a great deal of coordination is required to generate forecasts at the points of sale outside the company and with the production system since errors can generate capacity conflicts.

Likewise, such coordination and communication must also be with suppliers since they must deliver in small quantities and frequently, as identified in *PUS* programs. Likewise, Selçuk (2013) indicates that the best for the production system is to have large production batches, which prevents them from changing machines and tools, so a large batch indicates little flexibility and capacity to supply small production batches. The last paragraph indicates that if the company produces under a *PUS*, then it should be able to generate *SLP*, so the following hypothesis is proposed:

H_1. *Pull system* has a direct and positive impact on Small-lot production.

When producing under a *PUS*, frequently, more than one product must be produced daily, indicating the levels of flexibility in the production system (Gillam et al. 2018). However, it requires a very accurate forecasting system, preferably that production orders are fixed or with low variability about lot size (Jun and Ji 2016).

Fulfilling production orders generated in a *PUS* requires balanced production lines, which avoids bottlenecks and generalized delays in delivery times. If this balance does not exist, there may be problems associated with accumulating work in the production process, and specific machines or activities will become vital. They will be a great risk for the entire production system (Arumugam and Saral 2015).

Authors such as Morales Méndez and Rodriguez (2017) indicate that balancing production lines can be solved with simple algorithms. However, other tools such as TPM should be focused on those machines with limitations, and high availability must be guaranteed; or seeking a reduction of waste of time and movements in those critical points (Che Ani and Abdul Azid 2020). Thus, it is considered that a *PUS* system is related to the leveling of *UPL*, so the following hypothesis is proposed:

H_2. *Pull system* has a direct and positive effect on the *Uniform production level*.

To achieve *UPL* in a production system, it is preferable to have *SLP* and not large production batches. This way, errors can be identified quickly in the production system and favors managers to be more attentive to any generated changes (Niu et al. 2020). Some other authors indicate that, for an inexperienced manager, an *SLP* can be so fast that it is possible that he/she does not reach to identify the errors in the productive system, and they are a risk (França et al. 2019). Thus, *SLP*s require great knowledge about product quality characteristics (Kato et al. 2018).

However, *PUS* and *SLP* force companies to have a small number of products in inventory. It is accepted that forecasts always possess an error that is impossible to measure, generating variability (Logeshwaran et al. 2021). In addition, *SLP* also forces rigorous *Quality control* to ensure customer satisfaction, which requires managers again to be attentive to the conditions under which small and balanced batches are produced (Kolosowski et al. 2015). However, as Ishchenko and Fedotov (2004) mentioned, *SLP*-based *UPL* systems can only be obtained with a comprehensive training program at all organizational levels of the company. Therefore, the following hypothesis is proposed:

H₃. *Small-lot production* has a direct and positive effect on the Uniform Production level.

At this point, it is worth asking, what are the benefits of having a *PUS*, an *SLP*, or a *UPL* in a production system? The truth is that there are many, and the structural equation model presented here assumes that quality is one of them. It has been mentioned that *PUS* forces to produce what the customer requires so that production batches are generated accurately. There are no margins of error since otherwise, the production order would have to be inflated to generate an inventory that could be used to adjust for shrinkage. Hence, quality in the production system is vital (Lenarduzzi et al. 2021).

Similarly, other LMTs are required to help ensure quality, such as Kanban (Claudio and Krishnamurthy 2009) and maintenance of machines and tools (Zhou et al. 2019). Even authors such as Puchkova et al. (2016) mention that there must be a balance between push and pull systems to guarantee quality in products. Ge and Krishnamurthy (2011) indicate that it is necessary to have quality in the information generated along the supply chains since this is often not reliable and generates problems within the production systems. Quality in a product is one of the most critically affected characteristics.

In addition, since production batches are usually small in a *PUS*, then signals are required to improve communication within the production processes, which facilitates rapid decision making (Qasim and Al-Ani 2018). However, as He et al. (2015) mentioned, workers must have the authority to stop the production process if they detect an error (He et al. 2015) and thus prevent non-quality products from continuing the production line and reaching the customer.

H₄. *Pull system* has a direct and positive effect on Quality Control.

Also, since small production batches are produced, the quality control of the production process must be more demanding, as there is no room for errors. Quality managers and supervisors must be highly trained to detect quality deviations in products in real-time and take corrective actions immediately (Kolosowski et al. 2015). Specifically, Savino et al. (2008) have proposed a quality management system based on several indicators that allow monitoring quality when the emphasis is on customized products and thus manufactured in an *SLP*.

For their part, Deng et al. (2014) indicate that it is possible to have *SLP* and guarantee quality; however, given that quality sampling is with few parts, they recommend using non-parametric control charts based on moving averages. Given that *SLP* requires the delivery of raw materials in small batches, it is also required that suppliers be certified and highly reliable, which avoids acceptance problems in warehouses and reduces uncertainty in the supply process (Quigley et al. 2018). Thus,

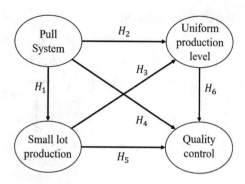

Fig. 5.1 Proposed model—pull system and quality control

SLP is considered to affect the quality delivered in a production process, and the following hypothesis is proposed:

H$_5$. *Small-lot production* has a direct and positive effect on *Quality control*.

UPL offers advantages over *QUC* since manufacturing more than one model per day requires production managers to have sufficient knowledge about the products, their technical specifications, and possible errors that could be generated, allowing them to take immediate action when deviations arise (Chua et al. 2003). Also, a *UPL* requires a similar production size from one day to another. It allows operators to have deep knowledge about the products and be experts in the processes and activities they perform. So, they can identify errors, which guarantees the quality of the production process (García-Alcaraz et al. 2016), so the following hypothesis is proposed.

H$_6$. Uniform lot production has a direct and positive effect on *Quality control*.

To better understand the relationships between the four latent variables analyzed in the model, Fig. 5.1 graphically illustrates their dependence, where *PUS* has been placed in the upper left and *QUC* in the lower right as a response to the different manufacturing practices performed.

5.4 Structural Equation Model Evaluation

Table 5.1 shows that all the latent variables analyzed in this model (*PUS, SLP, UPL* and *QUC*) have sufficient validity, so they were integrated into the structural equation model (although not with all their component items found in the questionnaire). The model's efficiency indexes are listed below:

- Average path coefficient (APC) = 0.286, $P < 0.001$.
- Average R-squared (ARS) = 0.256, $P < 0.001$.
- Average adjusted R-squared (AARS) = 0.250, $P < 0.001$.
- Average block VIF (AVIF) = 1.121, acceptable if ≤ 5, ideally ≤ 3.3.
- Average full collinearity VIF (AFVIF) = 1.519, acceptable if ≤ 5, ideally ≤ 3.3.

Fig. 5.2 Evaluated model

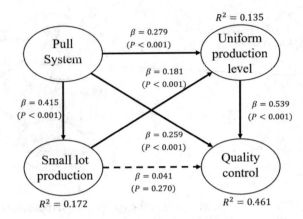

- Tenenhaus GoF (GoF) = 0.389, small ≥ 0.1, medium ≥ 0.25, large ≥ 0.36.

From the value of the above indices, it is concluded that there is sufficient dependence between the related variables, given that the APC is 0.286 and is statistically significant ($P < 0.001$). Likewise, the model has sufficient predictive validity in the latent dependent variables since the value of ARS and AARS is also statistically significant with a $P < 0.001$.

Also, it is observed that there are no collinearity problems between the latent variables since the AVIF and AFVIF indexes are lower than 3.3 (remember that some items have been eliminated because they did not contribute to explaining the latent variable variability). Finally, it is observed that the Tenenhaus GoF index is greater than 0.36, so it is concluded that the data in the sample fit the proposed SEM. Therefore, considering that the model complies with the efficiency indexes, we proceed to its interpretation.

Figure 5.2 presents the model evaluated. For each hypothesis or relationship between variables that have been established, a standardized β value and the associated p-value are indicated to measure its statistical significance. Likewise, an R^2 value is associated with each latent variable to measure the variability explained by the independent variables.

5.4.1   Direct Effects and Effect Sizes

To validate all proposed hypotheses in Fig. 5.1, direct effects are used. In general terms, it is observed in Fig. 5.2 that five of the six relationships between the variables analyzed are statistically significant, given that the p-values associated with the β parameters are less than 0.05 (significance level). In contrast, the relationship between ULP and *QUC* was not statistically significant and is illustrated in red.

Table 5.3 summarizes the hypotheses analyzed and the conclusions obtained based on the p-value associated with each β parameter. For example, for H$_1$, the *Pull system*

Table 5.3 Direct effects

H_i	Independent variable	Dependent variable	β	p-valor	Conclusion
H_1	PUS	SLP	0.415	< 0.001	Supported
H_2	PUS	UPL	0.279	< 0.001	Supported
H_3	SLP	UPL	0.181	= 0.003	Supported
H_4	PUS	QUC	0.259	< 0.001	supported
H_5	SLP	QUC	0.041	= 0.270	No supported
H_6	UPL	QUC	0.539	< 0.001	Supported

Table 5.4 Effect size for R^2

	PUS	SLP	UPL	R^2
SLP	0.172			0.172
UPL	0.090	0.045		0.135
QUC	0.113	0.010	0.338	0.461

has a direct and positive effect on *Small-lot production* since the associated p-value is less than 0.05, given that when the first variable increases its standard deviation by one unit, the second goes up by 0.438 units. In other words, producing only what the customer requires, and when the customer requires it, favors the existence of small production flows and orders.

Figure 5.2 shows that the latent dependent variables have an associated R^2 value, which indicates the percentage of variance explained by the independent variables that directly affect it. Table 5.4 summarizes each latent independent variable's contribution to the dependent variables in explaining their variance.

It is observed that *PUS* explains 0.172 of *SLP*, and its relationship was statistically significant, indicating that other variables explain *SLP* that were not analyzed in this model. *PUS* and *SLP* contribute to explain 0.172 of *UPL* (with 0.090 and 0.045, respectively), while *PUS*, *SLP* and *UPL* explain 0.461 of *QUC* (with 0.113, 0.010 and 0.338, respectively). The above indicates that *UPL* is the most important variable in explaining *QUC*, given the effect size; however, *SLP* contributes very little and almost negligibly in *QUC* since that relationship was statistically non-significant.

This assertion is confirmed by observing that the effect's most significant value is between *UPL* and *QUC*. This makes sense since always making identical production batches generates a greater knowledge of the technical specifications of the products by the operators and managers, who can make decisions quickly to avoid quality nonconformities, thus avoiding the rejection of production batches. If a product were manufactured sporadically, operators and managers would not have the same experience level as those produced daily.

Table 5.5 Sum of indirect effects

	PUS	SLP
UPL	0.075 ($p = 0.056$) ES = 0.024	
QUC	0.208 ($p < 0.001$) ES = 0.091	0.098 ($p = 0.019$) ES = 0.024

5.4.2   Sum of Indirect Effects

Figure 5.2 shows that the relationship between *SLP* and *QUC* is not statistically significant. However, *SLP* has an indirect effect on *QUC* through the mediating variable *UPL*, so it is convenient to make a complete analysis of the indirect effects that occur between the variables, which are illustrated in Table 5.5. In this case, the first row contains the independent variable (*PUS* and *SLP*) and the first column the dependent variables (*UPL* and *QUC*). For each relationship, a *p*-value and the size of the effect generated are indicated.

In this case, it is observed that the indirect effect between *PUS* and *UPL* is not statistically significant since the associated *p*-value is less than 0.05, where *SLP* acts as a mediating variable. Also, it is observed that the indirect effect from *PUS* to *QUC* has a value of 0.208, representing almost 80% of the direct effect. In other words, the *SLP* and *UPL* variables favor the relationship between *PUS* and *QUC* since the effect is positive.

5.4.3   Total Effects

The sum of direct and indirect effects on this occasion is vital since the direct effect between *SLP* and *QUC* was not statistically significant, as was the indirect effect between *PUS* and *UPL*, so the total effects are estimated and represented by the sum of the direct and indirect effects. Table 5.6 illustrates these total effects with a standardized β value, the associated *p*-value, and effect size (ES).

Table 5.6 Total effects

	PUS	SLP	UPL
SLP	$\beta = 0.415$ ($p < 0.001$) ES = 0.172		
UPL	$\beta = 0.354$ ($p < 0.001$) ES = 0.114	$\beta = 0.181$ ($p = 0.003$) ES = 0.045	
QUC	$\beta = 0.467$ ($p < 0.001$) ES = 0.204	$\beta = 0.139$ ($p = 0.018$) 0.034	$\beta = 0.539$ ($p < 0.001$) 0.338

It is important to note that all effects are statistically significant at the 95% confidence level. This indicates that the relationship of *SLP* with *QUC* does exist indirectly, and the total effect also exists; that is, small batches do help to improve the quality of the final product, but they must be uniform, produced consistently, and daily. In the same way, although the indirect effect between *PUS* and *UPL* was not statistically significant, the direct effect was, as well as the total effect. Producing under a pull policy in which only what is required is manufactured does not indirectly favor the generation of *UPL* through *SLP*. However, the relationship is direct and does not require a mediating variable.

5.4.4 Sensitivity Analysis

As in all models, a sensitivity analysis is performed to determine the effect of the independent variables' implementation levels (high or low) on the dependent variables. Scenarios are estimated, where probabilities for all latent variables are estimated in isolation for their low and high levels. When two variables are presented simultaneously or jointly, they are represented by &, finally, the conditional probability is reported and represented by IF. A low level of implementation in an LM practice occurs when the standardized z-value is less than -1, and a high level of implementation occurs when the standardized z-value is greater than 1.

Sensitivity analysis is critical because if the probability is interpreted to measure the risk of obtaining negative scenarios for such important variables as quality, it will help take immediate corrective actions. In the same way, one can observe which variables favor high occurrence levels and focus on executing those critical LM practices. Table 5.7 summarizes the calculated probabilities, where the "$-$" sign represents the probabilities for low implementation or execution levels of a variable, while the "$+$" sign represents the probabilities for high levels.

5.5 Conclusions and Industrial Implications

5.5.1 Conclusions from the Structural Equation Model

Using the information obtained in Fig. 5.2, which shows the model evaluated, the summary of information associated with the hypotheses inTable 5.3, and the indirect effects in Table 5.4 and total effects in Table 5.5, we conclude the following concerning the structural equation model.

H_1. There is sufficient statistical evidence to state with 95% confidence that the *Pull system* has a direct and positive impact on *Small-lot production* since when the first variable increases its standard deviation by one unit, the second increases it by 0.415 units. In other words, the implementation of a manufacturing practice

Table 5.7 Sensitivity analysis

			PUS		SLP		UPL	
			+	−	┤	−	+	−
			0.155	0.173	0.173	0.155	0.136	0.145
SLP	+	0.173	& = 0.055 If = 0.353	& = 0.009 If = 0.053				
	−	0.155	& = 0.018 If = 0.118	& = 0.064 If = 0.368				
UPL	+	0.136	& = 0.064 If = 0.412	& = 0.009 If = 0.053	& = 0.036 If = 0.211	& = 0.023 If = 0.147		
	−	0.145	& = 0.018 If = 0.118	& = 0.064 If = 0.368	& = 0.009 If = 0.053	& = 0.032 If = 0.206		
QUC	+	0.155	& = 0.055 If = 0.353	& = 0.014 If = 0.079	& = 0.050 If = 0.289	& = 0.014 If = 0.088	& = 0.064 If = 0.467	& = 0.009 If = 0.063
	−		& = 0.009 If = 0.059	& = 0.064 If = 0.368	& = 0.014 If = 0.079	& = 0.023 If = 0.147	& = 0.000 If = 0.000	& = 0.082 If = 0.563

based on *PUS* favors the execution of *SLP*. That is due because *PUS* is based on forecasts obtained daily from the field or points of sale and, production is scheduled according to what has been sold, consumed, or that the customer wishes to acquire and for which there is already a commercial contract. In addition, small but frequent batches are produced, making expert operators and managers relate to the characteristics of the products.

H_2. There is sufficient statistical evidence to state with 95% confidence that the *Pull system* directly affects *Uniform production level*. When the first variable increases its standard deviation by one unit, the second goes up by 0.279 units directly. However, there is also an indirect effect of 0.075, which gives a total effect of 0.354. The above indicates that implementing a *PUS* system favors the execution of the manufacturing practice called *UPL*, which may be based on daily production forecasts, manufacturing what is already sold, or a sales forecast in a short period. This ensures a decrease in storage costs of finished products, a greater flow of economic resources, improved relationships with customers and suppliers, and increased material turnover.

H_3. There is sufficient statistical evidence to state with 95% confidence that *Small-lot production* has a direct and positive effect on the Uniform Production level since when the first variable increases its standard deviation by one unit, the second increases it by 0.181 units. In other words, having an *SLP* system favors having a ULP with minor or very smoothed variations about the quantity to be produced.

H_4. There is sufficient statistical evidence to state with 95% confidence that the *Pull system* has a direct and positive effect on *Quality control* since when the first variable increases its standard deviation by one unit, the second goes up by 0.259 units directly. However, indirectly the effect is 0.208, which gives a total effect of 0.467. The industrial implication of this relationship is that it is necessary to produce what the client asks for based on short-period forecasts. This practice improves the quality of the products since a high level of specialization in the quality characteristics of the products is achieved, avoiding errors, waste, and wastage due to non-compliance with technical specifications.

H_5. There is not enough statistical evidence to declare with 95% confidence that *Small-lot production* has a direct and positive effect on *Quality control* since when the first variable increases its standard deviation by one unit, the second increases it by only 0.041 and has an associated *p*-value of 0.270. However, it has an indirect effect through ULP of 0.098, which results in a total effect of 0.139, which is statistically significant. *SLP* requires *ULP* to indirectly affect *QUC*, indicating that lot uniformity is a quality-assuring variable.

H_6. There is sufficient statistical evidence to state with 95% confidence that uniform lot production has a direct and positive effect on *Quality control* since when the first variable increases its standard deviation by one unit, the second increases it by 0.539 units in a di-linear manner. This may be because few changes are required in the production system since batches are manufactured frequently and in quantities almost always the same and slight variation.

It is essential to mention that, when reviewing the β parameters, it is observed that this relationship has the highest value, indicating the high relationship between these variables. This is verified when observing the effect size, which is 0.338 from a total of 0.461 (ULP explains 73.31% of the variance contained in *QUC*).

5.5.2 Conclusions from the Sensitivity Analysis

Conclusions are drawn based on the information contained in Table 5.7, and each of the relationships between variables for the four possible combinations of implementation scenarios is discussed. Recall that scenarios with high variable levels are denoted by "+" and scenarios with low levels are denoted by "−." Thus *PUS*+ will indicate a high level of a *Pull system*, for example.

For the relationship $PUS \rightarrow SLP$ (H_1), it is observed that *PUS*+ favors the execution of *SLP*+ with a probability of 0.353, although these two variables are unlikely to occur together. Likewise, it is observed that *PUS*+ is not related to *SLP*−, since the probability of occurrence is low, but if *PUS*− occurs, then there is a probability of 0.368 that *SLP*− will occur; furthermore, *PUS*− is not associated with *SLP*+, since the probability is almost zero. In re-summary, managers should strive for *PUS*+ levels to ensure proper execution of *SLP*+.

The *PUS* importance in the industry is reaffirmed by reviewing the relationship with *UPL* (H_2), where it is observed that *PUS*+ favors the proper execution of *UPL*+ with a probability of 0.412, which is a high value; moreover, it is unlikely that *PUS*+ is associated with *UPL*−, since the probability is low, only 0.064. It is also observed that the probability of *PUS*− generating *UPL*− is very high and represents a risk since the conditional probability is 0.368. Likewise, *PUS*− will not generate *UPL*+

since the probabilities have shallow values. In short, from an industrial point of view, this finding indicates that a *PUS* system passed in short-term forecasts helps to have *UPL*, so managers should observe if this production system favors them.

Also, *PUS*+ directly affects *QUC*+ and is observed in H₄, since the conditional probability for *QUC*+ given that *PUS*+ has occurred at its high level is 0.353, which indicates that a production system based on short-term forecasts favors the implementation of quality systems. However, if *PUS*+ occurs, there is a 0.059 probability of having a *QUC*−, which confirms this relationship. On the other hand, if *PUS*− occurs, then there is a high probability of not meeting quality expectations since the conditional probability is 0.368; furthermore, *PUS*− is unlikely to generate *QUC*−, variable conditional probability is only 0.079.

The above indicates that those managers should obtain *PUS*+ to ensure *SLP*+, *UPL*+, and *QUC*. This occurs because a *PUS*+ system requires a daily, constant, and precise demand to generate production. So that the production batches are small, uniform, and with little variability in quantity to be produced. This means that operators and managers are very familiar with the technical specifications of the products, know them in all their attributes, and can identify deviations quickly, which guarantees *QUC*.

Another meaningful relationship is *SLP* → *UPL* (H₃), which makes common sense, but requires statistical analysis. It is observed that *SLP*+ slightly favors the presence of *UPL*+ since the conditional probability is 0.211. However, *SLP*+ does not favor the presence of *UPL*−, since the probability is only 0.053; that is, managers who seek to establish *SLP* in their production lines are also favoring *UPL*. Likewise, it is observed that, even if *SLP*+ occurs, there is a conditional probability of 0.147 that *UPL*− occurs, indicating that there are many other factors that must be analyzed to interpret the occurrence of *UPL* at its high levels. Also, it is observed that *SLP*− favors the *UPL*− with a probability of 0.206, which is a risk for managers.

For the relationship *SLP* → *QUC* (H₅), it is observed that *SLP*+ favors *QUC*+ since the conditional probability that the second variable occurs at its high level, given that the first one has occurred, is 0.289. That indicates that managers who have *SLP* implemented in their production lines will obtain a higher quality in their products, which may be due to the high specialization of human resources. However, it is observed that *SLP*+ is only weakly associated with *QUC*− since the probability is only 0.079. On the other hand, if *SLP*− is present, there is only a conditional probability of 0.088 that *QUC*+ will occur. Finally, that *SLP*− can generate a *QUC*− with a probability of 0.147, representing a risk.

From the two previous paragraphs, in which *SLP* is related to *UPL* and *QUC*, it is observed that the production of small batches is of vital importance to be able to have level production orders and to guarantee the quality of the products. It is also concluded that low levels of *SLP* implementation represent a risk to produce uniform batches and quality.

Finally, in the relationship between *UPL* and *QUC*, it is observed that there are the highest conditional probabilities and β parameter values, which indicates the high relationship between both variables. It is observed that if *UPL*+ occurs, then *QUC*+ can occur with a probability of 0.457, and on this occasion, *UPL*+ is never associated with $QUC-$, since the probability is zero. However, if $UPL-$ occurs, then $QUC+$ can only occur at a probability of 0.063, which is very low but indicates that other variables are associated with product quality. Finally, if $UPL-$ occurs, there is a high risk of $QUC-$ occurring since the probability is 0.563.

Annex: Additional Data for Support Validation

T ratios for path coefficients

	PUS	*SLP*	*UPL*
SLP	6.638		
UPL	4.354	2.772	
QUC	4.022	0.615	8.833

Confidence intervals for path coefficients

	PUS		*SLP*		*QUC*	
	LCL	UCL	LCL	UCL	LCL	UCL
SLP	0.292	0.537				
UPL	0.153	0.404	0.053	0.309		
QUC	0.133	0.385	−0.090	0.172	0.420	0.659

T ratios for loadings

Item	*PUS*	*SLP*	*UPL*	*QUC*
PUS01	13.992			
PUS04	13.842			
PUS05	15.487			
PUS06	13.204			
SML01		12.566		
SLP02		12.801		
SLP03		12.692		
SLP404		14.603		
SLP05		12.136		
SLP06		11.919		
UPL02			12.610	
UPL03			14.165	

(continued)

(continued)

Item	PUS	SLP	UPL	QUC
UPL04			14.580	
UPL05			11.664	
UPL06			10.783	
QUC01				12.721
QUC02				12.781
QUC03				14.542
QUC04				11.894
QUC07				13.013
QUC08				11.462

Confidence intervals for loadings

	PUS		SLP		UPL		QUC	
	LCL	UCL	LCL	UCL	LCL	UCL	LCL	UCL
PUS01	0.699	0.927						
PUS04	0.691	0.919						
PUS05	0.775	1.000						
PUS06	0.658	0.887						
SML01			0.624	0.855				
SLP02			0.637	0.867				
SLP03			0.631	0.862				
SLP404			0.730	0.957				
SLP05			0.602	0.833				
SLP06			0.590	0.822				
UPL02					0.627	0.857		
UPL03					0.708	0.935		
UPL04					0.729	0.956		
UPL05					0.576	0.809		
UPL06					0.528	0.763		
QUC01							0.633	0.863
QUC02							0.636	0.866
QUC03							0.727	0.954
QUC04							0.589	0.821
QUC07							0.648	0.878
QUC08							0.565	0.799

PLSc reliabilities (Dijkstra's rho_a's)

PUS	SLP	UPL	QUC
0.839	0.854	0.867	0.875

Additional indices (indicator corr. matrix fit)

- Standardized root mean squared residual (SRMR) = 0.105, acceptable if ≤ 0.1
- Standardized mean absolute residual (SMAR) = 0.086, acceptable if ≤ 0.1
- Standardized chi-squared with 209 degrees of freedom (SChS) = 11.003, $P < 0.001$
- Standardized threshold difference count ratio (STDCR) = 0.948, acceptable if ≥ 0.7, ideally = 1
- Standardized threshold difference sum ratio (STDSR) = 0.860, acceptable if ≥ 0.7, ideally = 1.

Additional reliability coefficients

	PUS	SLP	UPL	QUC
Dijkstra's PLSc reliability	0.839	0.854	0.867	0.875
True composite reliability	0.892	0.886	0.866	0.885
Factor reliability	0.892	0.886	0.866	0.885

Correlations among l.vs. with sq. rts. of AVEs

	PUS	SLP	UPL	QUC
PUS	0.821	0.401	0.315	0.435
SLP	0.401	0.752	0.138	0.212
UPL	0.315	0.138	0.753	0.623
QUC	0.435	0.212	0.623	0.750

Full collinearity VIFs

PUS	SLP	UPL	QUC
1.412	1.194	1.642	1.829

HTMT ratios (good if < 0.90, best if < 0.85)

	PUS	SLP	UPL
SLP	0.479 ($p < 0.001$)		
UPL	0.387 ($p < 0.001$)	0.181 ($p < 0.001$)	
QUC	0.515 ($p < 0.001$)	0.252 ($p < 0.001$)	0.575 ($p < 0.001$)

References

S. Arumugam, J. Saral, Identification of bottleneck elements in cellular manufacturing problem. Procedia Comput. Sci. **74**, 181–185 (2015). https://doi.org/10.1016/j.procs.2015.12.097

M.N. Che Ani, I. Abdul Azid, Solving the production bottleneck through minimizing the waste of motion for manual assembly processes. Adv. Struct. Mater. **131** (2020). https://doi.org/10.1007/978-3-030-46036-5_17

L.Y. Chua, W.N. Shwe, S. Muto, Y. Odaka, A screen-mottling inspection and analysis system for CRT-manufacturing quality control. J. Soc. Inf. Display **11**(1 SPEC), 231–236 (2003). https://doi.org/10.1889/1.1831712

D. Claudio, A. Krishnamurthy, Kanban-based pull systems with advance demand information. Int. J. Prod. Res. **47**(12), 3139–3160 (2009). https://doi.org/10.1080/00207540701739589

Y.H. Deng, H.P. Zhu, G.J. Zhang, H. Yin, F.M. Liu, Nonparametric control charts design and analysis for small lot production based on the moving average. Adv. Mater. Res. **988** (2014). https://doi.org/10.4028/www.scientific.net/AMR.988.461

J. Ebner, P. Young, J. Geraghty, Intelligent self-designing production control strategy: dynamic allocation hybrid pull-type mechanism applicable to closed-loop supply chains. Comput. Ind. Eng. **135**, 1127–1144 (2019). https://doi.org/10.1016/j.cie.2019.04.005

J.W. Fowler, S.-H. Kim, D.L. Shunk, Design for customer responsiveness: decision support system for push–pull supply chains with multiple demand fulfillment points. Decis. Support Syst. **123**, 113071 (2019). https://doi.org/10.1016/j.dss.2019.113071

S.L.B. França, D.F. Dias, A.E.B. Freitag, O.L.G. Quelhas, M.J. Meiriño, Lean manufacturing application analysis in the inventory management of a furniture industry, in *Springer Proceedings in Mathematics and Statistics* (2019), pp. 207–219. https://doi.org/10.1007/978-3-030-14969-7_18

J.L. García-Alcaraz, W. Adarme-Jaimes, J. Blanco-Fernández, Impact of human resources on wine supply chain flexibility, quality, and economic performance. Ing. Investig. **36**(3), 74–81 (2016). https://doi.org/10.15446/ing.investig.v36n3.56091

D. Ge, A. Krishnamurthy, Pull systems with advance demand information: effect of information quality, in *61st Annual IIE Conference and Expo Proceedings* (2011)

S.L. Gillam, S. Holbrook, J. Mecham, D. Weller, Pull the andon rope on working memory capacity interventions until we know more. Lang. Speech Hear. Serv. Sch. **49**(3), 434–448 (2018). https://doi.org/10.1044/2018_LSHSS-17-0121

X. Guan, S. Ma, Z. Yin, The impact of hybrid push–pull contract in a decentralized assembly system. Omega **50**, 70–81 (2015). https://doi.org/10.1016/j.omega.2014.07.008

Y. He, H. Sun, K.K. Lai, Y. Chen, Organizational empowerment and service strategy in manufacturing. Serv. Bus. **9**(3), 445–462 (2015). https://doi.org/10.1007/s11628-014-0233-2

Y. Hoga, The uncertainty in extreme risk forecasts from covariate-augmented volatility models. Int. J. Forecast. **37**(2), 675–686 (2021). https://doi.org/10.1016/j.ijforecast.2020.08.009

V.A. Ishchenko, V.A. Fedotov, Engineering training desk-top complex for small-lot casting production. Litejnoe Proizv. **4**, 35–38 (2004)

Y. Jun, I. Ji, Demand-pull technology transfer and needs-articulation of users: a preliminary study. Procedia Comput. Sci. **91**, 287–295 (2016). https://doi.org/10.1016/j.procs.2016.07.079

O. Kabadurmus, M.B. Durmusoglu, Design of pull production control systems using axiomatic design principles. J. Manuf. Technol. Manag. **31**(3), 620–647 (2019). https://doi.org/10.1108/JMTM-07-2019-0272

C. Kato, N. Hiraiwa, T. Arai, J. Yanagimoto, Multi-station molding machine for attaining high productivity in small-lot productions. CIRP Ann. **67**(1), 293–296 (2018). https://doi.org/10.1016/j.cirp.2018.04.012

M. Kolosowski, J. Duda, J. Tomasiak, Statistical process control in conditions of piece and small lot production, in *Annals of DAAAM and Proceedings of the International DAAAM Symposium* (2015), pp. 147–155. https://doi.org/10.2507/26th.daaam.proceedings.021

V. Lenarduzzi, V. Nikkola, N. Saarimäki, D. Taibi, Does code quality affect pull request acceptance? An empirical study. J. Syst. Softw. **171** (2021). https://doi.org/10.1016/j.jss.2020.110806

J. Logeshwaran, R.M. Nachiappan, S. Nallusamy, Evaluation of overall manufacturing line effectiveness with inventory between sustainable processes in continuous product line manufacturing system. J. Green Eng. **11**(1), 104–121 (2021). https://doi.org/10.1108/17410380610688278

J.D. Morales Méndez, R.S. Rodriguez, Total productive maintenance (TPM) as a tool for improving productivity: a case study of application in the bottleneck of an auto-parts machining line. Int. J. Adv. Manuf. Technol. (2017). https://doi.org/10.1007/s00170-017-0052-4

J. Nelson, Pull versus push: lessons from lean manufacturing, Chap 3, in *Becoming a Lean Library* (Chandos Publishing, 2016), pp. 29–49. https://doi.org/10.1016/B978-1-84334-779-8.00003-3

B. Niu, Q. Li, Y. Liu, Conflict management in a multinational firm's production shifting decisions. Int. J. Prod. Econ. **230** (2020). https://doi.org/10.1016/j.ijpe.2020.107880

A. Puchkova, J. Le Romancer, D. McFarlane, Balancing push and pull strategies within the production system. IFAC-PapersOnLine 66–71 (2016). https://doi.org/10.1016/j.ifacol.2016.03.012

R.H. Qasim, M.S. Al-Ani, A quality control system for white circle pull using image analysis. J. Eng. Appl. Sci. **13**(14), 5738–5745 (2018). https://doi.org/10.3923/jeasci.2018.5738.5745

J. Quigley, L. Walls, G. Demirel, B.L. MacCarthy, M. Parsa, Supplier quality improvement: the value of information under uncertainty. Eur. J. Oper. Res. **264**(3), 932–947 (2018). https://doi.org/10.1016/j.ejor.2017.05.044

M.M. Savino, S. Apolloni, Y. Ouzrout, Product quality pointers for small lots production: a new driver for quality management systems. Int. J. Prod. Dev. **5**(1–2), 199–211 (2008). https://doi.org/10.1504/IJPD.2008.016378

B. Selçuk, Adaptive lead time quotation in a pull production system with lead time responsive demand. J. Manuf. Syst. **32**(1), 138–146 (2013). https://doi.org/10.1016/j.jmsy.2012.07.017

S.L. Takeda Berger, E.M. Frazzon, A.M. Carreirao Danielli, Pull-production system in a lean supply chain: a performance analysis utilizing the simulation-based optimization, in *2018 13th IEEE International Conference on Industry Applications, INDUSCON 2018—Proceedings* (2019), pp. 870–874. https://doi.org/10.1109/INDUSCON.2018.8627187

C. Zhang, D. Fang, X. Yang, X. Zhang, Push and pull strategies by component suppliers when OEMs can produce the component in-house: the roles of branding in a supply chain. Ind. Mark. Manage. **72**, 99–111 (2018). https://doi.org/10.1016/j.indmarman.2018.02.012

B. Zhou, G. Cheng, Z. Liu, Z. Liu, A preventive maintenance policy for a pull system with degradation and failures considering product quality. Proc. Inst. Mech. Eng. Part E J. Process Mech. Eng. **233**(2), 335–347 (2019). https://doi.org/10.1177/0954408918784414

Chapter 6
Model 3. Supplier Network and Inventory Minimization

Abstract This chapter presents a structural equation model (SEM) relating four latent variables associated with lean manufacturing (LM) practices: *Flexible resources (FLR)*, *Supplier networks (SUN)*, *Quality control (QUC)*, and *Inventory minimization (INMI)*. *FLR* serves as the independent variable and the others as dependent variables. The LM practices are related through six hypotheses evaluated with the help of WarpPLS v.7.0 software using the partial least squares technique with information from 228 responses to a questionnaire applied to the maquiladora industry sector in northern Mexico. A sensitivity analysis is reported for each proposed hypothesis with low and high occurrence scenarios for each relationship. The model results show that the most critical factor in minimizing inventories is the networks built with suppliers, followed by adequate quality control to avoid defects and waste due to poor quality, and finally, having *Flexible resources* within the companies. The sensitivity analysis shows that to optimize their inventories, managers effectively and administrators of the different departments in the maquiladoras must ensure that each of the activities within each of the latent variables is carried out correctly.

Keywords SEM · Flexible resources · Supplier networks · Quality control · Inventory minimization

6.1 Model Variables and Their Validation

Table 6.1 shows the validation indices for each latent variable shown in the model in Fig. 6.1, which consists of four variables: *Supplier networks (SUN)* as an independent variable, which had seven items before validation and five after validation. *Flexible resources (FLR)* with seven items at the beginning of the validation and five after validation. *Quality control (QUC)* with eight items at the beginning of the analysis and six at the end. *Inventory minimization (INMI)* with seven items at the beginning and five after analysis. The latter three as dependent variables. The 21 items in the variables are integrated into a model for evaluation.

Table 6.1 Latent variable validation

Items	FLR		SUN		QAC		INMI	
	7	5	7	5	8	6	7	5
R^2	0.267				0.447		0.449	
Adjusted R^2	0.264				0.442		0.441	
Composite realiability	0.891		0.905		0.885		0.950	
Cronbach's alpha	0.846		0.868		0.843		0.934	
Average variance extracted	0.622		0.657		0.562		0.794	
Full collinearity VIF	1.437		1.827		1.813		1.734	
Q-squared	0.280				0.445		0.449	

Fig. 6.1 Proposed model—network and inventory minimization

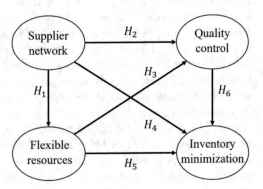

Table 6.1 shows the acronyms for each of the latent variables and the validation indices. The first row shows the items for each variable at the beginning and the end of the validation.

According to the R^2 and Adjusted R^2 indices, it is concluded that all the variables have sufficient predictive validity from a parametric point of view since the values are greater than 0.02. According to the composite reliability and Cronbach's alpha values, which are greater than 0.7, it is concluded that there is sufficient internal consistency in each of the variables. Concerning the Average variance extracted index values, it is concluded that the variables have sufficient convergent validity since the values are greater than 0.5. According to the full collinearity VIF values, it is concluded that there are no collinearity problems within the latent variables.

Finally, Q^2 represents the predictive validation from a non-parametric point of view. The values are greater than 0 and close to R^2, so it is concluded that there is sufficient predictive validity in each latent variable. According to the values of the indices illustrated in Table 6.1, it can be concluded that all the latent variables meet the validation criteria required to be integrated into a structural equation model and analyze the relevant relationships.

Annex shows some additional indices provided by the WarpPLS v.7.0 software to validate the variables and model evaluated.

6.2 Descriptive Analysis of the Items

Table 6.2 shows the descriptive analysis of the items that make up each variable. The table lists the items of each variable and associates them with a median value and an interquartile range. The mean is used to measure central tendency and the interquartile range as a dispersion measure.

As for the latent variable *flexible resources*, the highest value of the median is the one that corresponds to the fact that general-purpose machines are used in the companies to perform several essential functions. This means that, in general, there is agreement among the respondents on the use of flexible machines to perform various essential functions. The lowest median value is 4.10 and corresponds to the fact that workers can perform different types of work; that is, the respondents agree that the companies have a flexible workforce.

Concerning the items in the *Supplier networks* variable, high median values were obtained in all questions. The most important aspect in this variable is a consensus among the workers that the maquiladora companies strive to maintain a long-term relationship with their suppliers. This phenomenon is because the median value for this question is the highest (4.84) and with little dispersion (1.74). The activity or items that stand out the most is the agreement that the companies' suppliers deliver materials in the correct quantity and at the right time. The median value is 4.43, and the interquartile range is 1.84.

Concerning the latent variable *Quality control*, the participants agree that the maquiladora companies use visual aid systems as mechanisms to make problems visible since the median values reported are 5.16 and with very little dispersion in the data. In general, very high median values can be observed in this variable, which indicates that the companies monitor processes with statistical *Quality control* techniques, have teams of workers focused on quality, and meet regularly to discuss this topic. In addition, statistical techniques are used to reduce process variation.

Finally, for the values of the items in the latent variable *Inventory minimization*, it can be concluded that there is agreement among the participants about the reduction of raw material inventory since the median value is 5.09 and the responses have very little dispersion (1.66). This means that there is agreement that the finished product inventory is reduced (median of 5.04 and a dispersion of 1.77). This reduces the space required for storage with a median of 5.08 and a dispersion of 1.54.

Table 6.2 Descriptive analysis for items

	Median	Interquartile range
Flexible resources		
If a particular workstation has no demand, can production workers go elsewhere in the manufacturing facility to operate a workstation that has demand?	4.27	2.00
If one production worker is absent, another production worker can perform the same responsibilities?	4.22	2.13
Are production workers cross-trained to perform several different jobs?	4.27	2.19
We use general-purpose machines, which can perform several essential functions?	4.33	2.04
Are production workers capable of performing several different jobs?	4.10	2.14
Supplier networks		
We strive to establish long-term relationships with suppliers	4.84	1.74
We emphasize working together with suppliers for mutual benefits	4.82	1.65
We regularly solve problems jointly with suppliers	4.68	1.89
Development programs (such as engineering and quality management assistance) are provided to supplier	4.37	1.95
Our suppliers deliver materials to us just as it is needed (on just-in-time basis)	4.43	1.84
Quality control		
We use visual control systems (such as andon/line-stop alarm light, level indicator, warning signal, signboard) as a mechanism to make problems visible	5.16	1.84
Production processes on production floors are monitored with statistical quality control techniques	5.00	1.73
We have quality-focused teams that meet regularly to discuss quality issues	4.89	2.03
We use statistical techniques to reduce process variances	4.78	2.04
Quality problems can be traced to their source easily	4.57	2.09
Production workers are trained for quality control	4.51	2.03
Inventory minimization		
The work-in-process inventory (WIP) level has been significantly reduced	4.89	2.20
The level of raw material inventory has been significantly reduced	5.09	1.66
The inventory level of finished products has been significantly reduced	5.04	1.77
The general level of inventory has been significantly reduced	4.57	1.99
The storage space requirement has been significantly reduced	5.08	1.54

6.3 Hypotheses in the Model

The drive toward greater flexibility under LM management has forced manufacturers to outsource to suppliers who specialize in components and sub-assemblies and supply them in the required quantities and mix at the right time (Holweg 2007). When a supplier wants to produce and deliver specific parts for the customer under a contract, it usually has (at least in part) to redesign production processes. It may have to invest in new manufacturing equipment, tooling, and operator training and education (Bertrand 2003). Suppliers seeking to build closer links between themselves and critical customers will need to create a friendly environment in which employees are flexible. Also, suppliers must be prepared to make decisions without consulting management, be accountable, and communicate effectively with their counterparts in customer organizations (Kalwani and Narayandas 1995; Möller and Wilson 1995). Thus, the following hypothesis can be put forward:

H_1: *Supplier networks* have a direct and positive effect on *Flexible resources*

A *Quality control* system helps improve products and services. The system is continuously evaluated and modified to meet changing customer needs, improve productivity, reduce scrap and rework production, thus increasing usable products and reducing costs in the long run (Mitra 2016). Successful implementation of quality and continuous improvement programs in any manufacturing system directly relates to understanding the benefits of problem identification in supplier management (González et al. 2004). The correct choice of suppliers is a critical element for improving component quality (Li and Zabinsky 2011) since the quality of the components supplied allows a reduction of faults or waste during the manufacturing/assembly of the product and avoids after-sales problems (Paciarotti et al. 2014).

The collaborative customer–supplier relationship can be mutually beneficial if the supplier can learn from the customer and provide better quality and production control by adopting norms such as ISO and equivalent standards proactively and reactively to customer requirements (Ueki 2016). Based on the above, the following hypothesis can be put forward:

H_2: *Supplier networks* have a direct and positive effect on *Quality control*

Today's manufacturing is characterized by producing various products in small batches on the same set of machinery and equipment. This is why workers now have to perform larger jobs and perform a wide variety of tasks (Singh and Chauhan 2013) since labor flexibility is the ability to move from one job to another (Bobrowski and Park 1993). Increased employee skill acquisition can enable employees to conduct a quality inspection personally or give them access to productivity-related information (Murray and Gerhart 1998).

A multi-skilled, well-trained, and experienced workforce translates directly into higher productivity, better quality, and lower costs (Małachowski and Korytkowski 2016). A flexible workforce can produce a wide range of output assemblies and produce them with high efficiency (Chang 2004). A flexible workforce fits into a strategic emphasis on quality, and therefore, its training and coaching should be favored (Karuppan and Ganster 2004). Therefore, the following hypothesis can be put forward:

H_3: *Flexible resources* have a direct and positive effect on *Quality control*

A close relationship between buyer and supplier results in coordinated stock movements between the two. Reliable and timely delivery of products to the desired address, response to orders in a short time, and improved business performance (Lee et al. 2004). The competitive strategy developed by a close relationship with suppliers contributes to companies' production flexibility and better inventory control (Üstündağ and Ungan 2020). Due to the above, the following hypothesis can be put forward:

H_4: *Supplier networks* have a direct and positive effect on *Inventory minimization*

One of the fundamentals of LM is the multi-skilled employee, who has the competencies, experience, and knowledge to perform different simple and complex tasks (Małachowski and Korytkowski 2016). Labor flexibility and workstation flexibility help companies improve the operational performance of production systems through increased throughput, help reduce manufacturing flow times, and work-in-process (WIP) inventories. In the same way, they help to control inventory overages and improves their customer service while providing efficient use of both labor and equipment; it also helps in cycle time reduction (Bobrowski and Park 1993; Iravani 2010; Polakoff 1991; Rahman et al. 2013). In that sense, the following hypothesis is put forward:

H_5: *Flexible resources* has a direct and positive effect on *Inventory minimization*

The presence of defective items in raw material or finished goods inventories can affect SC coordination and, consequently, product flows between SC inventory levels can become unreliable (Roy et al. 2014). Using statistical *Quality control* tools such as Shewhart charts and histograms can help eliminate inventory waste (Leonov et al. 2020). Likewise, control charts can be used to measure performance and control inventory-related costs (Moisio and Virolainen 1993). In terms of inventory control, the effects of inventory reduction on productivity result from improved scheduling, maintenance, *Quality control*, and workforce management necessary for a just-in-time system to work (Mefford 1989). In that sense, quality system procedures support and even create the basis for effective inventory control (Moisio and Virolainen 1993). In that sense, the following hypothesis can be put forward:

H_6: *Quality control* has a direct and positive effect on *Inventory minimization*

The hypotheses described above are represented graphically in Fig. 6.1.

6.4   Evaluation of the Structural Equation Model

As mentioned above, the variables have been statistically validated using the indices shown in Table 6.1. Therefore, they can be integrated into a structural equation model to establish pertinent relationships. The proposed model is shown in Fig. 6.1, and before performing an analysis of the effects between the latent variables, the model must be validated. The validation is carried out using the quality and efficiency indexes shown below:

- Average path coefficient (APC) = 0.346, $P < 0.001$
- Average R-squared (ARS) = 0.388, $P < 0.001$
- Average adjusted R-squared (AARS) = 0.383, $P < 0.001$
- Average block VIF (AVIF) = 1.369, acceptable if ≤ 5, ideally ≤ 3.3
- Average full collinearity VIF (AFVIF) = 1.703, acceptable if ≤ 5, ideally ≤ 3.3
- Tenenhaus GoF (GoF) = 0.506, small ≥ 0.1, medium ≥ 0.25, large ≥ 0.36.

The dependence between the relationships (hypotheses) shown in Fig. 6.1 is tested using the APC index, and the predictive validity is tested using the ARS and AARS indices. Each hypothesis is tested with a confidence level of 95% and a significance level of (p-value) 5%. For each relationship, a β value is established, a hypothesis test where the null hypothesis $H_0 : \beta = 0$ and the alternative hypothesis $H_1 : \beta \neq 0$.

In that sense, p-values less than 0.05 allow us to reject the null hypothesis, indicating a relationship between the latent variables and that this relationship is of the size of β. In other words, there is a relationship between the two latent variables. In this case, the mean values and their p-values for each of the APC (0.346, $P < 0.001$), ARS (0.388, $P < 0.001$), and AARS (0.383, $P < 0.001$) indices are less than 0.05. So it can be observed that these relationships are statistically significant, and there is predictive validity between each of the relationships between the variables.

Concerning the values of VIF and AFVIF less than 3.3, it can be concluded that there are no collinearity problems between the latent variables. Finally, concerning the GoF index value of 0.506, the model has high explanatory power, and that the data fit the model.

Figure 6.2 shows the evaluated model where for each hypothesis, a p-value is shown. In addition, an R^2 value is shown, which reflects the prediction or explained variance of the independent latent variables over the dependent variables.

6.4.1   Direct Effects and Effect Sizes

As mentioned in the previous section, a hypothesis was established for each relationship between the latent variables. Figure 6.2 shows the values of β or the direct effect between the two variables and the p-value for each relationship. According to the p-values shown in Fig. 6.2, five of the six hypotheses stated are statistically significant as their p-values are less than 0.05. The relationship between *flexible resources* and

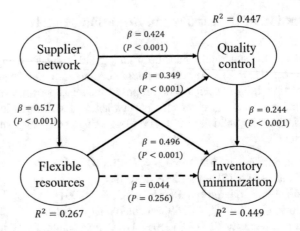

Fig. 6.2 Evaluated model

Inventory minimization is not, as its p-value $= 0.256$. Table 6.3 shows the summary of the β and p-values for each of the hypotheses and the conclusion for these.

Figure 6.2 shows that the largest β value or effect is between *Supplier networks* and *flexible resources* with a value of $\beta = 0.517$. This means that when the *Supplier networks* variable increases its standard deviation by one unit, the *flexible resources* variable increases its standard deviation by 0.517 units. In other words, *Supplier networks* are essential for resource flexibility within firms because if firms establish long-term relationships with their suppliers, they emphasize working together with them in achieving mutual benefits and problem solving if suppliers deliver materials as requested. If the firm provides development programs to suppliers, resource flexibility will be obtained, both in workers and machines.

Table 6.3 illustrates a summary of the results in Fig. 6.2 as a synthesis.

For each latent dependent variable in Fig. 6.2, an R^2 value is provided and indicates the variance explained by the independent variables. For example, in the case of *flexible resources*, which has a value of $R^2 = 0.267$, it indicates that the *Supplier networks* explain 26.7% of the variance. In the case of *Quality control*, its variance or $R^2 = 0.447$ is explained in 0.252% by *Supplier networks* and 0.195% by *flexible*

Table 6.3 Direct effects

H_i	Independent variable	Dependent variable	β	p-value	Conclusion
H_1	SUN	FLR	0.517	< 0.001	Supported
H_2	SUN	QUC	0.424	< 0.001	Supported
H_3	FLR	QUC	0.349	< 0.001	Supported
H_4	SUN	INMI	0.496	< 0.001	supported
H_5	FLR	INMI	0.044	$= 0.256$	No supported
H_6	QUC	INMI	0.244	< 0.001	Supported

Table 6.4 Effect size for R^2

	FLR	SUN	QUC	R^2
FLR		0.267		0.267
QUC	0.195	0.252		0.447
INMI	0.014	0.312	0.123	0.449

Table 6.5 Sum of indirect effects

	FLR	SUN
QUC		0.181 ($p < 0.001$) ES = 0.108
INMI	0.085 ($p = 0.036$) ES = 0.027	0.170 ($p = 0.005$) ES = 0.107

resources. In this case, the variable that contributes most to *QUC* is *SUN*. Finally, the R^2 of *Inventory minimization* is explained by 0.014 for *FLR*, 0.312 for *SUN*, and 0.123 for *QUC* for a total of 0.449. Concerning this last variable, it is known that the hypothesis of *FLR* with *INMI* was not statistically significant. Therefore, the contribution of the R^2 is minimal; that is, it does not contribute statistically to explain *INMI*. Therefore, it can be concluded that resource flexibility does not contribute directly to *Inventory minimization* (Table 6.4).

6.4.2 Sum of Indirect Effects

Another of the effects analyzed in structural equation models is the indirect or moderating effects, which occur through third variables. In the model evaluated in Fig. 6.2, there are three indirect effects. The one from *FLR* to *INMI* through *SUN* and *QUC* with a $\beta = 0.085$ and a p-value $= 0.036$. In this case, this effect is statistically significant, but it is small; *FLR* contributes very little indirectly to *INMI*. In addition, *SUN* has an indirect effect on *QUC* through *FLR* of $\beta = 0.181$ with a p-value < 0.05; i.e., *SUN* contributes indirectly to *QUC* through *FLR*. Finally, *SUN* has an indirect effect on *INMI* through *FLR* and *QUC* of $\beta = 0.170$ with a p-value < 0.05 and explains 0.107 of the variance of *INMI* (Table 6.5).

6.4.3 Total Effects

The total effects are the sum of the direct effects and the indirect effects. Table 6.6 summarizes the six total effects presented in the evaluated model. It can be seen that all effects are statistically significant at p-values < 0.05. In the case of the direct effect that was not statistically significant (*FLR* on *INMI*), the indirect effect is added for a

Table 6.6 Total effects

	FLR	SUN	QUC
FLR		$\beta = 0.517$ ($p < 0.001$) ES = 0.267	
QUC	$\beta = 0.349$ ($p < 0.001$) ES = 0.195	$\beta = 0.604$ ($p < 0.001$) ES = 0.360	
INMI	$\beta = 0.129$ ($p = 0.026$) ES = 0.041	$\beta = 0.666$ ($p < 0.001$) ES = 0.419	$\beta = 0.244$ ($p < 0.001$) ES = 0.123

value of $\beta = 0.129$ with a p-value $= 0.026$, making it a statistically significant total effect. It is undoubtedly a somewhat low β value, but in the end, it is concluded that FLR does contribute to explaining INMI.

It is important to note that the variable that contributes most to INMI is SUN, with a value of $\beta = 0.666$. SUN is the variable that contributes most to each independent variable since the value of β on QUC is 0.604 and on FLR is 0.517. This indicates the importance of suppliers for the companies since they contribute to QUC, FLR, and INMI in the maquiladora companies.

6.4.4 Sensitivity Analysis

Table 6.7 presents the sensitivity analysis for each of the hypotheses proposed in the model in Fig. 6.1. It shows the probabilities that activities performed within each latent variable will be executed correctly or in a high scenario (+) or, on the contrary,

Table 6.7 Sensitivity analysis

			SUN		FLR		QUC	
			+	−	+	−	+	−
			0.145	0.132	0.164	0.164	0.155	0.164
FLR	+	0.164	& = 0.045 If = 0.313	& = 0.027 If = 0.207				
	−	0.164	& = 0.005 If = 0.031	& = 0.059 If = 0.448				
QUC	+	0.155	& = 0.064 If = 0.438	& = 0.005 If = 0.034	& = 0.059 If = 0.361	& = 0.005 If = 0.028		
	−	0.164	& = 0.000 If = 0.000	& = 0.059 If = 0.448	& = 0.005 If = 0.028	& = 0.077 If = 0.472		
INMI	+	0.123	& = 0.055 If = 0.375	& = 0.000 If = 0.000	& = 0.059 If = 0.361	& = 0.005 If = 0.028	& = 0.064 If = 0.412	& = 0.005 If = 0.028
	−	0.164	& = 0.009 If = 0.063	& = 0.068 If = 0.517	& = 0.027 If = 0.167	& = 0.036 If = 0.222	& = 0.005 If = 0.029	& = 0.050 If = 0.306

that these activities will not be performed properly or in a low scenario (−). The joint probabilities (&) are also presented; that is, the occurrence of both variables in the different scenarios (+ +, + −, + −, − +, − −). In addition, the conditional probability (If) is presented, i.e., that one variable occurs in a specific scenario, given that the other occurs in a specific scenario. For example, the probability that the dependent variable occurs in its high (+) scenario given that the independent variable occurs in its high (+) scenario is shown.

The probability of performing activities adequately in each variable is 0.145, 0.164, 0.155, and 0.123 for *SUN*, *FLR*, *QUC*, and *INMI*, respectively. Conversely, the probabilities that the activities for each variable are not performed adequately are 0.132, 0.164, 0.164, and 0.164 for *SUN*, *FLR*, *QUC*, and *INMI*, respectively.

It is important to note that both variables' joint probabilities of high and low scenarios are small. For example, for hypothesis H_1, the probability of finding the variable *SUN+* & *FLR+* is 0.045. On the contrary, the probability of *SUN−* & *FLR−* is 0.059. This means that it is improbable that the activities will be performed correctly or incorrectly together, and managers and engineers must focus on that activity.

In the case of the conditional probability for hypothesis H_1, the probability of *FLR* occurring in its high scenario, given that *SUN* occurs in its high scenario, is 0.313. Conversely, the probability of *FLR* occurring in its low scenario, given that *SUN* occurs in its low scenario, is 0.448. In conclusion, for activities to have *flexible resources* within the companies, administrators or managers must establish long-term relationships with their suppliers. In addition, they should emphasize working together with them to achieve mutual benefits and problem solving and deliver the materials as requested. However, the company must provide supplier development programs. Otherwise, it will not have *flexible resources*. The conclusions for the other hypotheses are presented in the conclusions section.

6.5 Conclusions and Industrial Implications

The conclusions of both the structural equation model and the sensitivity analysis are shown below. Some industrial applications are also presented, according to the Mexican maquiladora industry sector.

6.5.1 Conclusions of the Model

Regarding findings obtained from the model in Fig. 6.2, it is possible to conclude the following (H_1 has been explained as an example above and here is excluded from the analysis):

H_2: There is enough statistical evidence to declare that *Supplier networks* have a direct and positive effect on *Quality control* because when the first variable increases its standard deviation in one unit, the second one goes up by 0.424 units and can explain its 0.252 of variance. In other words, it can be concluded that *Supplier networks* contribute through long-term relationships with pro-suppliers, collaboration with them to achieve mutual benefits, through joint problem solving. Also, these suppliers must deliver materials just-in-time, which will help establish *Quality control* programs within the companies. In addition, *Supplier networks* indirectly affect *Quality control* through *flexible resources* with a $\beta = 0.181$ and with an effect size of ES $= 0.108$. *Supplier networks* help firms have *flexible resources*, which, in turn, help to have *Quality control* in the maquiladoras. Therefore, *Supplier networks* have a total effect on *Quality control* of $\beta = 0.604$ with an effect size ES $= 0.360$.

H_3: There is enough statistical evidence to declare that *flexible resources* directly and positively affect *quality control* because when the first variable increases its standard deviation in one unit, the second one goes up by $\beta = 0.349$ units with an effect size of ES $= 0.195$. This means that flexible human resources and machinery will help companies have better *Quality control* directly on the production lines.

H_4: There is enough statistical evidence to declare that *Supplier networks* directly and positively affect *Inventory minimization* because when the first variable increases its standard deviation in one unit, the second one goes up by 0.496 units. In other words, through long-term relationships, *Supplier networks* collaborate to achieve mutual benefits through joint problem solving. These suppliers deliver materials in time to help minimize work in process, raw material, and finished product inventories. Similarly, *Supplier networks* contribute to *Inventory minimization* indirectly through *flexible resources* and *Quality control* with $\beta = 0.170$ and an effect size of ES $= 0.107$. The total effect of *Supplier networks* on *Inventory minimization* is $\beta = 0.666$ and an ES $= 0.419$.

H_5: There is not enough statistical evidence to declare that *flexible resources* directly and positively affect *Inventory minimization* because the p-value associated is 0.256, greater than the maximum allowed 0.05. However, indirectly there is an effect, but it is small, and it is through *Quality control*. In other words, *flexible resources* help establish Quality control in the companies, which, in turn, helps *Inventory minimization*.

H_6: There is enough statistical evidence to declare that *Quality control* directly and positively affects *Inventory minimization* because when the first variable increases its standard deviation in one unit, the second one goes up by 0.244 units with an effect size ES $= 0.123$. In other words, *Quality control* through visual control systems, monitoring through statistical techniques of production processes, through quality teams focused on discussing quality problems, through the use of statistical techniques to reduce process variability helps *Inventory minimization* within industries. The above is concluded since rejections due to poor product quality are avoided if there is quality control. Consequently, inventories of work in process, raw materials, and finished products are reduced.

6.5.2   Conclusions of the Sensitivity Analysis

From the sensitivity analysis in Table 6.7, we conclude the following (the analysis for H_1 is omitted since it has already been explained above as an example). Remember that in this case, the sign "+" indicates a high-level scenario, and a "−" sign indicates a level scenario:

H_2: There is a probability of 0.064 that $SUN+$ and $QUC+$ occur together, i.e., it is improbable that *Quality control* activities and *Supplier networks* are simultaneously established in their high scenarios. In contrast, there is a probability of 0.059 that these variables are jointly present at their low levels. For both cases, the probabilities are small, and it is possible that not occur.

However, here the conditional probabilities are important since there is a probability of 0.438 that QUC will occur at its high level given that SUN will occur in its high-level scenario. In other words, to establish good *Quality controls* within companies, managers need to ensure that they establish long-term relationships with their suppliers. Managers must emphasize working with them to achieve mutual benefits and problem solving and deliver the materials as requested. In addition, the company should provide supplier development programs. On the other hand, if administrators or managers do not establish *Supplier networks*, there is a 0.448 probability of not having reasonable *Quality control* within the company.

H_3: The probability that FLR and QUC occur together in the high scenario is 0.059 and that they occur together in a low scenario is 0.077. In other words, it is improbable that *Quality control* will occur at its high level and that *flexible resources* will occur at its high level together. Also, it is improbable that $QUC-$ and FLR− occur together. This means that there improbable to find a high degree of flexibility in labor, machinery, and equipment with good quality.

However, the conditional probability of finding $QUC+$, given that $FLR+$ has occurred, is 0.361, indicating that $FL+$ favors the $QUC+$ in production lines. So managers must be focused on flexibility regarding machinery, human resources, and management.

H_4: There is a 0.055 probability of SUN+ and INMI+ co-occurring, which is an opportunity area for managers, as this probability should be higher. However, if SUN+ occurs, there is a conditional probability of 0.375 of INMI+ occurring, demonstrating the strong relationship between those two variables.

Furthermore, managers can be confident that their investments to achieve SUN+ will yield a return since SUN− is not associated with INMI+. In other words, SUN+ never generates INMI−. Finally, it is essential to mention that SUN− can be detrimental to INMI−, since there is a conditional probability of occurrence of 0.517. In other words, low levels of SUN also lead to low levels of INMI.

H_5: The probability of finding FLR+ and INMI+ co-occurring simultaneously is only 0.059, which is too low and unacceptable for managers, as these values are expected to be high. However, the conditional probability of INMI+ occurring, given that FLR+ has occurred, is 0.361, indicating its high dependence. Furthermore, it is unlikely that INMI− is found, given that FLR+ has occurred, since the conditional probability is only 0.167, indicating that other variables favor INMI − and not only FLR.

Furthermore, FLR− and INMI+ are improbable to co-occur, as they have a joint probability of only 0.005, but the probability of INMI− occurring, given that FLR− has occurred, is 0.222, which represents a risk for managers.

H_6: The probability of QUC+ and INMI+ co-occurring is only 0.064, which is very low. However, the conditional probability of INMI+ occurring, given that QUC+ has occurred, is 0.412, indicating the high relationship between those variables. Furthermore, the investments that managers make in obtaining QUC+ will be reflected in the fact that it is improbable to be associated with INMI− jointly or to occur as a consequence of it, as the probabilities are 0.005 and 0.029, respectively.

Similarly, if QUC− occurs, it is improbable that INMI+ would co-occur, reinforcing the above statement. However, managers should be alert to these levels of QUC−, as it can generate INMI− with a probability of 0.306, severely affecting inventories throughout the supply chain.

Annex: Additional Data for Support Validation

T ratios for path coefficients

	FLR	SUN
FLR		8.434
QUC	5.521	6.789
INMI	0.657	8.061

Confidence intervals for path coefficients

	FLR		SUN		QUC	
	LCL	UCL	LCL	UCL	LCL	UCL
FLR			0.397	0.637		
QUC	0.225	0.473	0.301	0.546		
INMI	−0.087	0.175	0.376	0.617	0.117	0.370

T ratios for path coefficients for loadings

	FLR	SUN	QUC	INMI
FLR01	12.446			
FLR02	14.926			
FLR03	14.656			
FLR04	12.282			
FLR05	13.146			
SUN02		14.279		
SUN03		15.526		
SUN04		14.179		
SUN05		12.728		
SUN07		12.932		
QUC01			12.720	
QUC02			12.781	
QUC03			14.542	
QUC04			11.895	
QUC07			13.013	
QUC08			11.463	
INMI01				15.606
INMI02				16.026
INMI03				15.947
INMI04				16.445
INMI05				13.672

Confidence intervals for loadings

	FLR		SUN		QUC		INMI	
	LCL	UCL	LCL	UCL	LCL	UCL	LCL	UCL
FLR01	0.618	0.849						
FLR02	0.747	0.973						
FLR03	0.733	0.959						
FLR04	0.609	0.841						
FLR05	0.655	0.884						
SUN02			0.714	0.941				
SUN03			0.777	1.002				
SUN04			0.709	0.936				
SUN05			0.633	0.863				
SUN07			0.644	0.874				

(continued)

(continued)

	FLR		SUN		QUC		INMI	
	LCL	UCL	LCL	UCL	LCL	UCL	LCL	UCL
QUC01					0.633	0.863		
QUC02					0.636	0.866		
QUC03					0.727	0.954		
QUC04					0.589	0.821		
QUC07					0.648	0.878		
Q UC08					0.565	0.799		
INMI01							0.781	1.005
INMI02							0.802	1.026
INMI03							0.798	1.022
INMI04							0.823	1.046
INMI05							0.682	0.911

PLSc reliabilities (Dijkstra's rho_a's)

FLR	SUN	QUC	INMI
0.862	0.873	0.869	0.935

Additional indices (indicator corr. matrix fit)

- Standardized root mean squared residual (SRMR) $= 0.101$, acceptable if ≤ 0.1
- Standardized mean absolute residual (SMAR) $= 0.081$, acceptable if ≤ 0.1
- Standardized chi-squared with 209 degrees of freedom (SChS) $= 6.895, P < 0.001$
- Standardized threshold difference count ratio (STDCR) $= 0.952$, acceptable if $\geq$ 0.7, ideally $= 1$
- Standardized threshold difference sum ratio (STDSR) $= 0.864$, acceptable if $\geq$ 0.7, ideally $= 1$.

Additional reliability coefficients

	FLR	SUN	QUC	INMI
Dijkstra's PLSc reliability	0.862	0.873	0.869	0.935
True composite reliability	0.891	0.905	0.885	0.950
Factor reliability	0.891	0.905	0.885	0.950

Correlations among l.vs. with sq. rts. of AVEs

	FLR	SUN	QUC	INMI
FLR	0.789	0.321	0.549	0.306
SUN	0.321	0.811	0.515	0.625
QUC	0.549	0.515	0.750	0.474
INMI	0.306	0.625	0.474	0.891

Full collinearity VIFs

FLR	SUN	QUC	INMI
1.437	1.827	1.813	1.734

HTMT ratios (good if < 0.90, best if < 0.85)

	FLR	SUN	QUC
SUN	0.369 ($p < 0.001$)		
QUC	0.649 ($p < 0.001$)	0.599 ($p < 0.001$)	
INMI	0.345 ($p < 0.001$)	0.697 ($p < 0.001$)	0.536 ($p < 0.001$)

References

J.W.M. Bertrand, Supply chain design: flexibility considerations, in *Handbooks in Operations Research and Management Science*, vol. 11 (2003), pp. 133–198

P.M. Bobrowski, P.S. Park, An evaluation of labor assignment rules when workers are not perfectly interchangeable. J. Oper. Manag. **11**(3), 257–268 (1993)

A. Chang, On the measurement of labor flexibility. Paper presented at the 2004 IEEE International Engineering Management Conference (IEEE Cat. No. 04CH37574) (2004)

M.E. González, G. Quesada, C.A. Mora Monge, Determining the importance of the supplier selection process in manufacturing: a case study. Int. J. Phys. Distrib. Logist. Manag. **34**(6), 492–504 (2004). https://doi.org/10.1108/09600030410548550

M. Holweg, The genealogy of lean production. J. Oper. Manag. **25**(2), 420–437 (2007)

S.M. Iravani, Design and control principles of flexible workforce in manufacturing systems, in *Wiley Encyclopedia of Operations Research and Management Science* (2010)

M.U. Kalwani, N. Narayandas, Long-term manufacturer-supplier relationships: do they pay off for supplier firms? J. Mark. **59**(1), 1–16 (1995)

C.M. Karuppan, D.C. Ganster, The labor–machine dyad and its influence on mix flexibility. J. Oper. Manag. **22**(6), 533–556 (2004). https://doi.org/10.1016/j.jom.2004.08.005

H.L. Lee, V. Padmanabhan, S. Whang, Information distortion in a supply chain: the bullwhip effect. Manage. Sci. **43**(4), 546–558 (2004)

O.A. Leonov, N.Z. Shkaruba, Y.G. Vergazova, P.V. Golinitskiy, U.Y. Antonova, Quality control in the machining of cylinder liners at repair enterprises. Russ. Eng. Res. **40**(9), 726–731 (2020). https://doi.org/10.3103/S1068798X20090105

L. Li, Z.B. Zabinsky, Incorporating uncertainty into a supplier selection problem. Int. J. Prod. Econ. **134**(2), 344–356 (2011)

B. Małachowski, P. Korytkowski, Competence-based performance model of multi-skilled workers. Comput. Ind. Eng. **91**, 165–177 (2016). https://doi.org/10.1016/j.cie.2015.11.018

R.N. Mefford, The productivity nexus of new inventory and quality control techniques. Eng. Costs Prod. Econ. **17**(1), 21–28 (1989). https://doi.org/10.1016/0167-188X(89)90051-7

A. Mitra, *Fundamentals of Quality Control and Improvement* (Wiley, 2016)

U. Moisio, V.-M. Virolainen, Inventory control and quality system requirements—a case study. Int. J. Prod. Econ. **30–31**, 415–425 (1993). https://doi.org/10.1016/0925-5273(93)90109-X

K.K. Möller, D.T. Wilson, *Business Marketing: An Interaction and Network Perspective* (Springer Science & Business Media, 1995)

B. Murray, B. Gerhart, An empirical analysis of a skill-based pay program and plant performance outcomes. Acad. Manag. J. **41**(1), 68–78 (1998). https://doi.org/10.5465/256898

C. Paciarotti, G. Mazzuto, D. D'Ettorre, A revised FMEA application to the quality control management. Int. J. Qual. Reliab. Manag. **31**(7), 788–810 (2014). https://doi.org/10.1108/IJQRM-02-2013-0028

J. Polakoff, Reducing manufacturing costs by reducing cycle time. Corp. Control. **4**(2), 62–64 (1991)

N.A.A. Rahman, S.M. Sharif, M.M. Esa, Lean manufacturing case study with kanban system implementation. Procedia Econ. Financ. **7**, 174–180 (2013). https://doi.org/10.1016/S2212-567 1(13)00232-3

M.D. Roy, S.S. Sana, K. Chaudhuri, An economic production lot size model for defective items with stochastic demand, backlogging and rework. IMA J. Manag. Math. **25**(2), 159–183 (2014)

T.P. Singh, G. Chauhan, Significant parameters of labour flexibility contributing to lean manufacturing. Glob. J. Flex. Syst. Manag. **14**(2), 93–105 (2013). https://doi.org/10.1007/s40171-013-0033-x

Y. Ueki, Customer pressure, customer–manufacturer–supplier relationships, and quality control performance. J. Bus. Res. **69**(6), 2233–2238 (2016). https://doi.org/10.1016/j.jbusres.2015.12.035

A. Üstündağ, M.C. Ungan, Supplier flexibility and performance: an empirical research. Bus. Process Manag. J. **26**(7), 1851–1870 (2020). https://doi.org/10.1108/BPMJ-01-2019-0027

Chapter 7
Model 4. Integrative Model

Abstract This chapter presents a structural equation model (SEM) integrating the ten manufacturing practices that have been previously analyzed, which is of second-order. It is assumed that there are only four latent variables, which are *Machinery (MAC)*, *Production planning (PRP)*, and *Production process (PRR)*, which are independent and second-order, and explain *Quality control (QUC)* as first order and dependent variable. The variables are related to six hypotheses, and the model is evaluated using the partial least squares technique with information from 220 responses to a survey applied to the Mexican manufacturing sector. Findings indicate that for the *Production process*, the most critical variable is the *Machinery* available for the execution of the activities. Also, to achieve *Quality control*, the most critical variable is the *Production process*, followed by *Machinery*. *Production planning* has a low direct effect on the *Production process* and *Quality control*, although it is statistically significant.

Keywords SEM · Machinery · Production planning · Production process · Quality control

7.1 Model Variables and Their Validation

This model comprises three independent variables (*Machinery, Production planning,* and *Production process*) and one dependent variable (*Quality control*). In this case, it is assumed that the company accepts production orders based on identifying the production capacity it has installed through its *Machinery*. That is why this variable occupies the upper left-hand side in the model, on which all the other variables depend. Based on this installed capacity, the company begins to carry out *Production planning*, which is then executed through the *Production process* variable. Combining these three variables results in product quality, the variable located in the lower right part as the dependent variable.

In this model, the latent independent variables are of second-order, and Table 7.1 indicates how they are integrated by the LM practices that have been analyzed in other models.

© The Author(s), under exclusive license to Springer Nature Switzerland AG 2022   97
J. R. Díaz-Reza et al., *Best Practices in Lean Manufacturing*,
SpringerBriefs in Applied Sciences and Technology,
https://doi.org/10.1007/978-3-030-97752-8_7

Table 7.1 Second-order variables integration

Second-order variable	Latent variables as items
Machinery (MAC)	Cellular layout (*CEL*)
	Total productive maintenance (*TPM*)
	Single minute exchange of die (*SMED*)
Production planning (PRP)	Push system (*PUS*)
	Small lot production (*SLP*)
	Supplier networks (*SUN*)
Production process (PRR)	Inventory minimization (*INMI*)
	Uniform production level (*UPL*)
	Flexible resources (*FLR*)

Table 7.2 Latent variables validation

	QUC		*PRP*		*MAC*		*PRR*	
Items	6	4	3	3	3	3	3	3
R-squared	0.603		0.271				0.579	
Adjusted *R*-squared	0.598		0.267				0.575	
Composite realiability	0.885		0.827		0.848		0.817	
Cronbach's alpha	0.843		0.686		0.732		0.663	
Average variance extracted	0.562		0.616		0.651		0.600	
Full collinearity VIF	2.460		1.448		2.669		2.681	
Q-squared	0.604		0.278				0.583	

Table 7.2 illustrates the validation process of the latent variables analyzed in the model. The first row represents the number of items they had before and after the validation process. In this case, the second-order variables retained all the items (variables associated with manufacturing practices) since the number at the beginning and end of the validation process is the same; however, in the *QUC* variable, changes were observed in the validation process and the number of items.

In general terms, it is observed that the four second-order latent variables analyzed comply with the proposed validation indexes established in the methodology described in Chap. 3. For example, it is observed that R^2 and adjusted R^2 are greater than 0.02, so it is concluded that there is sufficient parametric predictive validity. Concerning Q^2, the values are similar to those of R^2, so it is concluded that there is sufficient non-parametric predictive validity.

Likewise, concerning Cronbach's alpha and Composite reliability indices, it is observed that the values are greater than 0.6, and it is concluded that there is sufficient content validity and internal validity for an exploratory analysis such as this case. Also, the values in the average variance extracted are greater than 0.5 in all variables, so it is concluded that there is sufficient convergent validity. Finally, the full collinearity index (VIF) is lower than 3.3 in all variables, indicating the absence of collinearity problems.

Please refer to Annex at the end of this chapter for readers interested in more validation indices for this model. The additional indexes are the T ratios for path coefficients, confidence intervals for path coefficients, T ratios for loadings, confidence intervals for loadings, PLSc reliabilities (Dijkstra's rho_a's), other indices (indicator correlation matrix fit), additional reliability coefficients, correlations among latent variables with square roots of AVEs, full collinearity VIFs and, HTMT ratios.

After the above analysis, the latent variables have sufficient validity and are integrated into the SEM. In this case, since the descriptive analysis for all manufacturing practices has already been reported previous models, they are omitted in this integrative model. If the reader wishes to see the descriptive analysis of the items for *Cell layout, Total productive maintenance, Single Minute Exchange of Die*, and *Inventory minimization*, please see model 1 in Chap. 4. For the descriptive analysis of the items contained in the *Pull system, Small lot production, Uniform production level*, and *Quality control*, please see model 2 in Chap. 5.

Thus, since the descriptive analysis of the items is omitted, the hypotheses are defined below in this integrative model.

7.2 Hypotheses in the Model

This model assumes that companies initiate *Production planning* based on their installed capacities, according to the *Machinery* available, and support programs, such as *TPM* and *SMED*. After the planning is done, they go to a *Production process*, where the plans and programs are executed on the machines, resulting in a quality product. There are four variables in the model, and they are related using six hypotheses, which are justified below.

Production planning starts by identifying the layout distribution in production lines, which allows identifying whether the requested products are feasible to be manufactured (Badiane et al. 2016). This is for analyzing the machines' capacities and the performance conditions in which they are found, the frequency and efficiency of maintenance they receive, and the calibration level (Xiang and Feng 2021).

Also, it is common to plan *SLP* since it helps to have more customized products for specific customers (Anzanello and Fogliatto 2011). *SLP* should be an industrial practice as much as possible (Chen and Tseng 2007) since it indicates productive flexibility and reputation for the company (García-Alcaraz et al. 2020). However, achieving such *SLP* requires rapid changes into the production system in machines and tools, which is supported by LM tools such as *SMED* (Tekin et al. 2019), although programs such as *TPM* also help ensure machine availability levels (Díaz-Reza et al. 2018).

Suppose the conditions inside the plant are favorable and a *CEL* allows the flow of production (Méndez-Vázquez and Nembhard 2019). In that case, the search for suppliers outside the company capable of supplying raw materials on time begins.

Currently, there are many multi-criteria techniques to select these suppliers. However, in a *PUS* system, purchasing managers must consider that it is vital to have the ability to offer small-scale and intermittent deliveries, this being one of the most important criteria (Schramm et al. 2020). In addition, the supplier must have remarkable resilience to adapt to unexpected changes that arise in the needs of the productive process and customers (Valipour Parkouhi et al. 2019).

Therefore, it is considered that the technological capabilities associated with LM *Machinery* and tools that are installed and the tools that are available for their maintenance affect *Production planning*, so the following hypothesis is proposed:

H_1. Lean manufacturing practices associated with *Machinery* have a direct and positive effect on LM practices in *Production planning*.

With the production plans already made, the same is executed on the production lines. That is, the machines and tools are put into operation, and here we seek to meet specific requirements, such as minimization of in-process inventory, which are indicative of problems associated with bottleneck and line balancing that reduce material flow efficiency (Vijay and Gomathi Prabha 2020; Adnan et al. 2016). Other LM tools that help to have a better flow of the production system are *TPM* and *SMED*. They increase machine availability levels, eliminating those bottlenecks due to machine breakdowns (Schindlerová et al. 2020).

Suppose the production system can make quick changes through *SMED*. In that case, it is possible to have a *UPL*, which requires a high commitment from the sales department to generate smoothed forecasts. There is not much variability concerning the quantity produced (Ilyina et al. 2017; Kennedy 2003). If that production system remains smooth, with no bottlenecks due to lack of production line balancing, then an *INMI* is obtained in all tasks, with no accumulation of raw materials waiting to be processed (Nawanir et al. 2021; Sharma and Singh 2012).

However, it is also important that operators can be multifunctional since this avoids bottlenecks. Suppose that absenteeism of an operator on a binding machine is reported, anyone else can occupy his or her work area and thus maintain the continuous flow of the production system producción (Sanderson et al. 2013). In addition, Suija-Markova et al. (2020) indicate that multifunctional employees have a greater impact on organizational performance, while Crews and Russ (2020) indicate that performing multiple activities represents a highly skilled multi-skilled operator-valued in a production system.

Therefore, in this book, it is considered that the LM practices focused on maintaining and preserving the *Machinery* in operation directly affect the practices executed in the *Production process*, and the following hypothesis is proposed:

H_2. LM practices associated with *Machinery* directly and positively affect the practices applied in the *Production process*.

Undoubtedly, the *PRP* is performed to obtain some benefit, and one of them is the reduction of uncertainty in the *PRR*. Thus, applying a *PUS* to production lines will be reflected in *UPL*, which has little variability from one day to the next, since only what has been sold in the previous day or in a concise time horizon is produced

(Fowler et al. 2019). In addition, if they already have a sales contract for a production lot, those products do not spend much time in storage and thus have *INMI* attending a *PUS* (Zou and Feng 2019; Nelson 2016).

This *INMI* is also greatly favored by the planned *SLPs* (Kim and Lee 2011) because these orders are often customized and manufactured for a specific customer and are generally already sold and will not spend much time in the warehouse. This *INMI* represents a cost reduction for the company, which would be greater if investments associated with storage and handling of materials had to be added (Zhang and Janet 2020).

It is also important to mention that when a *PUS* and *SLP* are planned, higher *FLR* is required; that is, operators should be able to perform several activities within the production line and be knowledgeable about the processes of different products (König et al. 2010; Hannan et al. 2019). In this case, high specialization of a *Production process* is beneficial only when having mass production, with extensive storage systems, although as mentioned by Zhang et al. (2019), it is possible to have a customized approach and mass production time.

From the above, it is observed that *PRP* are directly related to the execution of these in a productive system during PRR, so the following hypothesis is proposed:

H$_3$. LM practices associated with *Production planning* have a direct and positive effect on the *Production process*.

The application of LM practices offers many benefits. In this model, it is assumed that one of them is *QUC*, which can have several origins, and one of them is due to LM practices integrated into the *Machinery* variable (Singh and Ahuja 2014). If the machines are calibrated adequately due to a *TPM* program, then there are no deviations from the technical specifications of the products. On the other hand, if there is no proper *TPM* program, the machines will have low efficiency due to many defects identified in the products (Ahmad et al. 2019; Turcu et al. 2016). In summary, machines with proper maintenance will always offer better quality than those without, and de Deac (2010) and Deac (2011) reports are a clear example.

Another source of *QUC* is *CEL* since, as Hashemoghli et al. (2019) indicate, workers who work on only one group of products and activities are more skilled and make fewer mistakes in the *Production process*, which translates into better quality. Böllhoff et al. (2016) make a study in which they relate human error that affects product quality in a *CEL* and relate it to product quality and confirm that operators focused on only specific products and processes make fewer errors. However, it is essential to mention that this relationship is not new, as Seifoddini and Djassemi (1996) and Reynolds (1998) indicated more than 20 years ago.

A third LM tool or practice is *SMED*, which favors product quality and *QUC* in general since that is one of its benefits, as Gathen (2004) indicated when asking what this tool can do in a production system. Moreover, since lead time plays a vital role in quality, *SMED* is one of the most supportive tools to meet that goal, as quick changes avoid delays and non-compliance (Yazıcı et al. 2020). Currently, *SMED* has yielded results in the manufacturing sector, but its application has expanded. The reported benefits have been similar, continually decreasing setup times, as in the pharmaceutical industry (Karam et al. 2018).

From the above, it is observed that LM practices and tools applied to machines and tools affect *QUC*, so the following hypothesis is proposed:

H₄. The LM practices applied to *Machinery* have a direct and positive effect on *Quality control.*

For authors such as Osorio Gómez et al. (2018) and Salimian et al. (2020), product quality and product *Quality control* start with the raw material sourcing process. If these are of quality, the product will be reliable, assuming that the *Production process* is performed correctly. However, poor quality raw materials will always cause problems in the *Production process* and delays due to stoppages because specific components do not meet the required specifications (Yousef et al. 2015). Therefore, Lin et al. (2016) consider that supplier-manufacturer relationships must be solid and beyond simple sourcing. The supplier must be considered an external partner to the manufacturer and share the responsibility for the quality offered to the final customer (Awasthi 2015).

Another LM tool that favors *QUC* is *PUS* since only what is needed is produced. In *PUS*, there are smoothed variations in quantities to be produced, which allows managers and operators to adapt to changes, know the products in-depth, and, therefore, there is less possibility of making mistakes in the production lines and decreasing the quality of the production product (Fowler et al. 2019). However, Zhang and Janet (2020) indicate that under *PUS*, batches are small, and the *SLP* is met, requiring high operator knowledge levels. Nevertheless, in small batches, quality verification must be constant, as there is a minimal margin for error, given that production runs are short and in a short time.

Kolosowski et al. (2015) indicate that statistical process control in an *SLP* system must be extensively monitored and lists several conditions that must be met to execute it. In the same way, Savino et al. (2008) indicate that *SLP* requires specialized personnel who know the product and indicates some conditions to be met without affecting quality, such as low absenteeism, continuous support of *TPM* and *SMED*, among others. Kristoffersen (2014) points out that if these conditions are not met, there is a risk of affecting quality, on-time deliveries to the customer, and, perhaps, the loss of the customer.

For the above mentioned, it is observed that *Production planning* is related to *Quality control*, and the following hypothesis is proposed:

H₅. LM practices associated with *Production planning* have a direct and positive effect on *Quality control.*

However, it is worth mentioning that quality requires *Production planning*, but it is built in the *Production process* since many of the technical specifications are acquired there. For example, a program that offers an *INMI* will undoubtedly facilitate the flow of materials throughout the production system. However, it also facilitates on-time deliveries, fulfillment of complete production orders, and *Quality control* (Nawanir et al. 2021). The relationship between inventories and quality is ancient

Fig. 7.1 Proposed model

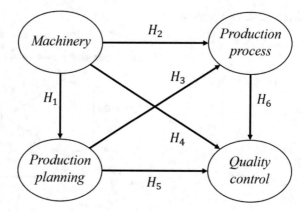

since Chung (1987) more than thirty years ago pointed out that high inventory levels in the *Production process* represent a lack of *Quality control*.

Other authors have proposed techniques and methodologies focused on *INMI*, but that simultaneously allows the application of an adequate *QUC*, such as Ernst et al. (1993) and Gasparini et al. (2009), highlighting the relationship between both variables.

Another LM practice that facilitates *QUC* is undoubtedly human resources, their skills, abilities, and the flexibility obtained from them. For example, García-Alcaraz et al. (2018) indicate the role of human resources and their skills in creating six sigma programs that guarantee and facilitate *QUC*. That role of operator cross-functionality and its effect on quality has also been studied in sectors other than industry, such as education (Gil et al. 2018; Alfonso and Mara 2016; Rus et al. 2014), supply chain (Zhang et al. 2016; Ojha et al. 2016), among others.

Considering that LM practices associated with the *Production process* affect quality execution and its control, the following hypothesis is proposed:

H_6. LM practices associated with the *Production process* have a direct and positive effect on *Quality control*.

The relationship of the latent variables established as hypotheses in this integrative model is illustrated in Fig. 7.1.

7.3 Structural Equation Model Evaluation

In Table 7.2, the validation indices of the latent variables analyzed have been reported. Since all variables have met the cut-off values, they have been integrated into the structural equation model. So, now the following validation indices are for the model as a whole and already integrated. The indices are as follows:

- Average path coefficient (APC) = 0.370, $P < 0.001$

Fig. 7.2 Evaluated model

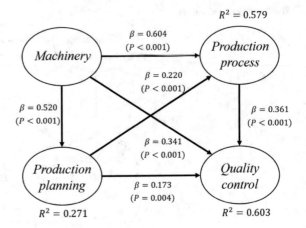

- Average R-squared (ARS) $= 0.484$, $P < 0.001$
- Average adjusted R-squared (AARS) $= 0.480$, $P < 0.001$
- Average block VIF (AVIF) $= 1.956$, acceptable if ≤ 5, ideally ≤ 3.3
- Average full collinearity VIF (AFVIF) $= 2.314$, acceptable if ≤ 5, ideally ≤ 3.3
- Tenenhaus GoF (GoF) $= 0.542$, small ≥ 0.1, medium ≥ 0.25, large ≥ 0.36.

From the above values, the relationships on average are statistically significant since the p-value associated with the APC is less than 0.001, indicating that inferences can be made with 99% confidence. Similarly, the values of ARS and AARS are greater than 0.02, and their associated p-value is lower than 0.05, so it is concluded that there is sufficient predictive validity in the model.

Regarding collinearity, it is observed that values for AVIF and AFVIF are lower than 3.3, so it is concluded that there are no collinearity problems between the latent variables of the integrative model. Finally, the value of Go is 0.542, higher than 0.36, and it is concluded that the data obtained from the sample fit the model adequately. Therefore, it is concluded that the model can be interpreted and is illustrated in Fig. 7.2.

Figure 7.2 illustrates a standardized β value and the associated p-value to measure its statistical significance. Likewise, an R^2 value is indicated for the latent dependent variables to measure the variance explained by the independent latent variables.

7.3.1   Direct Effects and Effect Size

The direct effects are used to validate the hypotheses proposed in Fig. 7.1. According to the p-values associated with the β indices, it is concluded that all the effects are statistically significant. Table 7.3 summarizes the conclusions for each proposed hypothesis, where the last column indicates the conclusion based on the hypothesis stated. For example, for hypothesis H_1, relationship $MAC \rightarrow PRP$, it is observed that

Table 7.3 Direct effects

H_i	Independent variable	Dependent variable	β	p-value	Conclusion
H_1	MAC	PRP	0.520	< 0.001	Supported
H_2	MAC	PRR	0.604	< 0.001	Supported
H_3	PRP	PRR	0.220	< 0.001	Supported
H_4	MAC	QUC	0.341	< 0.001	Supported
H_5	PRP	QUC	0.173	= 0.004	Supported
H_6	PRR	QUC	0.361	< 0.001	Supported

Table 7.4 Effect size for R^2

	PRP	MAC	PRR	R^2
QUC	0.102	0.243	0.258	0.603
PRP		0.271		0.271
PRR	0.131	0.448		0.579

when MAC increases its standard deviation by one unit, PRP increases it by 0.520 units. Furthermore, MAC can explain up to 27.1% of PRP, which indicates its high relationship. The other relationships will be explained in detail in the conclusions section.

Figure 7.2 illustrates an R^2 value for the latent dependent variables. However, more than one independent latent variable is explained, so Table 7.4 illustrates the effect sizes. The first row indicates the independent variables, while the first co-column indicates the dependent variables. The intersection of the variables represents the effect size of the independent variable on the dependent variable. In this case, MAC can explain up to 27.1% of PRP variability and is easy to estimate since it is the only independent variable that explains it. However, variables like QUC are explained by three independent variables.

Table 7.4 shows that three variables explain QUC (the other three variables of the model), and this is because they have a direct effect on it, explaining 60.3% of its variability ($R^2 = 0.603$); however, in this case, PRP explains only 10.2%, MAC 24.3% and finally, PRR explains 25.8%. From the above, it can be concluded that the most critical variables for obtaining QUC are MAC and PRR since together, they explain up to 50.1% (24.3% of MAC + 25.8% PRR). So, managers must focus on those variables if they want to guarantee QUC.

In PRP, this is explained only by MAC in 27.1%, and no other variable directly affects it in this model. Finally, PRR is explained 57.9% by MAC with 44.8% and PRP with 13.1%, indicating that MAC is an essential variable in PRP since it has the highest explanatory power, which makes sense, but is statistically significant demonstrated in this study.

If a manager wants to establish a critical path with these variables in the model, based on the indexes and standardized parameter values β, it will look as follows:

Table 7.5 Sum of indirect effects

	PRP	*MAC*
QUC	0.079 ($p = 0.046$) ES = 0.047	0.350 ($p < 0.001$) ES = 0.249
PRR		0.114 ($p = 0.046$) ES = 0.085

$MAC \rightarrow PRR \rightarrow QUC$, where for the relationship between $MAC \rightarrow PRR$ the β is 0.604, and for the relationship of $PRR \rightarrow QUC$ the β is 0.361. Although *PRP* is essential because it integrates the planning process, it is observed that this variable is not vital to guarantee *Quality control* according to the proposed model. However, a different result can be reported in another industrial sector or another geographical context.

7.3.2 Sum of Indirect Effects

Figures 7.1 and 7.2 show that some latent dependent variables have indirect effects on other latent independent variables through mediating variables. For example, *MAC* is an independent variable in the upper left of the model, and no variable explains it, but it has effects on all the others, some of them indirectly (Table 7.5).

Interesting conclusions emerge from the analysis of the indirect effects. Table 7.2 indicates that the direct effect between $MAC \rightarrow QUC$ was only 0.341 and was statistically significant and with a high value. However, the indirect effect was only 0.350, higher than the direct effect, indicating that the *PRP* and *PRR* variables are vital for *MAC* to strengthen the effect with *QUC*.

7.3.3 Total Effects

The direct and indirect effects are added together to determine the total effect from one variable to another. Table 7.6 illustrates this sum of total effects. Note that some of the total effects are equal to the direct effects because there are no indirect effects between these variables. Of particular interest are some total effects in this model. For example, the relationship $MAC \rightarrow QUC$, since the effect is 0.341 and the total effect is 0.690, which indicates that those two variables are primarily related, since, if machines are not out of calibration, then the quality cannot be guaranteed, which shows the importance of *TPM* in *QUC*.

Another total effect of particular interest is in the relationship between $MAC \rightarrow PRR$, which has a value of 0.719 and is the highest in Table 7.6. The importance of that relationship lies in the fact that it is in *PRR* where LM tools or practices associated with *TPM* and *SMED* are applied, which help to ensure better quality.

Table 7.6 Total effects

	PRP	MAC	PRR
QUC	$\beta = 0.253$ ($p < 0.001$) ES = 0.149	$\beta = 0.690$ ($p < 0.001$) ES = 0.492	$\beta = 0.361$ ($p < 0.001$) ES = 0.258
PRP		$\beta = 0.520$ ($p < 0.001$) ES = 0.271	
PRR	$\beta = 0.220$ ($p < 0.001$) ES = 0.131	$\beta = 0.719$ ($p < 0.001$) ES = 0.533	

This value increases this relationship because *PRP* is a mediating effect, which is statistically significant.

7.3.4 Sensitivity Analysis

It is of great interest for managers to know different scenarios, given certain conditions of the variables in the model. In this integrative model, as in the previous ones, a sensitivity analysis is presented in which an LM tool or practice is considered to have a low level of implementation if its standardized z-value is less than -1 $P(Z < -1)$. It is considered a high level of implementation if its standardized z-value is greater than 1 $P(Z > 1)$.

In this case, high levels are represented by the "+" sign and low levels by the "−" sign. The probabilities of finding the variables independently at their high and low levels, the probabilities of finding two latent variables jointly at their high and low levels, as well as their combinations, are analyzed, which is represented by "&". Finally, the conditional probabilities of a latent dependent variable occurring at its high or low level, given that an independent latent variable has occurred at its high or low level, are analyzed and represented by "IF".

Table 7.7 illustrates the sensitivity analysis. The independent latent variables with their two levels (high and low) are indicated by column. The dependent variables and their two levels (high and low) are indicated in the first column.

7.4 Conclusions and Industrial Implications

This section presents the conclusions, and industrial implications reached based on the findings obtained in the integrative model, emphasizing two main aspects:

- The structural equation model
- The sensitivity analysis.

Table 7.7 Sensitivity analysis

			MAC		PRP		PRR	
			+	−	+	−	+	−
			0.168	0.177	0.136	0.168	0.155	0.164
PRP	+	0.136	& = 0.068 If = 0.405	& = 0.000 If = 0.000				
	−	0.118	& = 0.000 If = 0.000	& = 0.045 If = 0.256				
PRR	+	0.136	& = 0.077 If = 0.459	& = 0.005 If = 0.026	& = 0.059 If = 0.433	& = 0.000 If = 0.000		
	−	0.168	& = 0.000 If = 0.000	& = 0.105 If = 0.590	& = 0.000 If = 0.000	& = 0.059 If = 0.500		
QUC	+	0.155	& = 0.064 If = 0.378	& = 0.005 If = 0.026	& = 0.055 If = 0.400	& = 0.000 If = 0.000	& = 0.073 If = 0.533	& = 0.005 If = 0.027
	−	0.164	& = 0.000 If = 0.000	& = 0.105 If = 0.590	& = 0.005 If = 0.033	& = 0.059 If = 0.500	& = 0.000 If = 0.000	& = 0.100 If = 0.595

7.4.1 Conclusions from the Structural Equation Model

For the conclusions based on the structural equation model, we use the information in Fig. 7.2 and Tables 7.3, 7.4, 7.5, and 7.6 associated with the direct effects, the effect sizes, the sum of indirect effects, and the total effects. For each of the hypotheses posed, we conclude the following:

H_1. There is sufficient statistical evidence to state with 95% confidence that lean manufacturing practices associated with *Machinery* have a direct and positive effect on *Production planning* practices since when the first variable increases its standard deviation by one unit, the second increases it by 0.520 units. This relationship indicates that tools such as *TPM, SMED*, and *CEL* should be considered before implementing production lines based on *PUS*. Those LM practices help determine the installed plant capacity, know the possible failures of the equipment, the bottlenecks that could be generated, and the route of the materials. In the same way, *SMED* helps determine the capacity of a production system based on *SLP*, where machine and tool changes are required constantly.

H_2. There is sufficient statistical evidence to state with 95% confidence that the LM practices associated with *Machinery* have a direct and positive effect on the practices applied in the *Production process* since when the first variable increases its standard deviation by one unit, the second increases it by 0.604 units (it is the highest direct effect). However, there is an indirect effect through *Production planning* of 0.114 units, which gives a total effect of 0.719 units (the highest in the model). This relationship indicates that these two variables have the highest relationship because *PRR* is where the tools associated with *MAC*, such as *SMED* and *TPM*, are executed to maintain a smooth production system. *CEL* facilitates having a system based on *UPL* and *SLP*, where production is uniform or with smoothed variations.

In addition, *SMED* and *TPM* allow an *INMI* since there are no broken machines and the flow of products is continuous. Therefore, they do not accumulate or remain as raw material in the process. Also, *SMED* or performing quick changeovers facilitates the achievement of *INMI*. At the same time, *CEL* enables operators to have specialized skills and abilities, which favors *FLR* that are multifunctional and perform diverse tasks.

H_3. There is sufficient statistical evidence to state with 95% confidence that LM practices associated with *Production planning* directly and positively affect practices associated with the *Production process* since when the first variable increases its standard deviation by one unit, the second increases it by 0.220 units. This relationship confirms that planning a *PUS* facilitates a *UPL* since production is planned in a brief period. The changes concerning the quantity to be produced are relatively similar or with few variations. In the same way, *SLP* generates that *INMI* can be implemented since production orders are small.

H_4. There is sufficient statistical evidence to state with 95% confidence that the LM practices applied to *Machinery* have a direct and positive effect on *Quality control* since when the first variable increases its standard deviation by one unit, the second increases it by 0.341 units. However, *MAC* also has an indirect effect on *QUC* through *PRP* and *PRR* (two- and three-segment) with a value of 0.350, giving a total effect of 0.690. The above indicates that a *CEL*-based distribution facilitates *QUC* since employees are specialists in some activities and can quickly identify deviations from technical specifications. Likewise, *SMED* facilitates *QUC* since quick changes are made and delivery times can be met. Finally, well-calibrated machines through *TPM* programs do not generate defects, which increases the quality of finished products.

H_5. There is sufficient statistical evidence to state with 95% confidence that LM practices associated with *Production planning* have a direct and positive effect on *Quality control* since when the first variable increases its standard deviation by one unit, the second increases it by 0.173 units. However, *PRP* also has an indirect effect on *QUC* through *PRR* of 0.079 units, giving a total effect of 0.253, both statistically significant. In other words, tools such as *PUS*, *SLP*, and *SUN* facilitate the implementation of *QUC*. For example, *PUS*, a production system based on short-term forecasts and demand, makes operators and managers familiar with the *QUC* system.

H_6. There is enough statistical evidence to state with 95% confidence that LM practices associated with the *Production process* directly and positively affect *Quality control* since when the first variable increases its standard deviation by one unit, the second increases it by 0.361 units. This relationship indicates that LM tools such as *UPL*, *INMI*, and *FLR* facilitate *QUC* activities. The implementation of *UPL* benefits *QUC* since approximately the same quantity of the same product is constantly produced, and operators and managers become experts in the product's technical specifications. *INMI* and *FLR* facilitate the implementation of *QUC* by reducing the amount of raw material in the process, facilitating on-time deliveries and less waste, and operators can perform multiple activities or operate different machines.

Other information of particular interest is shown in Table 7.6 concerning the sizes of the effects that exist in the relationships. For example, it is observed that the largest effect size is the one that *MAC* has on *PRR*, and that arises from a direct effect, with a value of 0.533, indicating that *MAC* can explain up to 53.3% of *PRR*. Another interesting relationship is that *MAC* and *QUC* have an effect size of 0.492, indicating that *QUC* is explained by *MAC* up to 49.2%, which confirms the importance of programs associated with *TPM* and *SMED* in the quality of the final product and its control.

7.4.2   Conclusions from the Sensitivity Analysis

One of the benefits of the analysis performed with the WarpPLS 7.0 software is that it allows estimating probabilities of specific scenarios since it allows managers and decision-makers to know the risks of unfavorable events. The following interpretation is based on the information in Table 7.7. In this case, the acronyms for the different variables are used, and, in addition, the sign "+" or "−" is added to indicate high and low levels, respectively. Thus, for example, *MAC+* refers to *Machinery* being at a high level of implementation, or *QUC−* refers to *Quality control* being low.

This sensitivity analysis complements the effect sizes reported when calculating direct, indirect, and total effects, where the measure is the variance explained. However, in the sensitivity analysis, relationships are noted on a probability basis.

For the relationship between $MAC \rightarrow PRP$ in H_1 (the highest relationship among variables), it is observed that *MAC+* favors *PRP+* with a probability of 0.405, so managers should focus on applying LM tools associated with *CEL*, *SMED*, and *TPM* to ensure that tools such as *PUS*, *SLP* and *SUN* can be implemented. In addition, it is observed that *MAC+* is not associated with *PRP−*, since the conditional probability is zero, which is verified by reviewing that also *MAC−* is not associated with *PRP−*. However, managers must consider that *MAC−* can generate a *PRP−* with a probability of 0.256, which can be interpreted as a risk.

In the relationship of *MAC* with *PRR* in H_2, it is observed that *MAC+* favors *PRR+* since the conditional probability is 0.459. Furthermore, it is observed that *MAC+* is not associated with *PRR−*, since the conditional probability is zero, which is verified by observing that *MAC−* is also not associated with *PRR+* where the probability exists but is a value close to zero. Finally, it is observed that if *MAC−* is present, then there is a high risk of obtaining *PRR−* since the conditional probability is 0.590. In other words, managers should seek to have *MAC+*, where *TPM*, *CEL*, and *SMED* support the production system to maintain high levels of machine availability.

The importance of *PRR* is proven when analyzing the relationship, it has with *PRP* in H_3. For example, if *PRP+* is present, a conditional probability of 0.433 that *PRR* will occur. However, it is noted that *PRP+* is not associated with *PRR−* or that *PRP−* is not associated with *PRP+* since the conditional probabilities are zero. However, if *PRP−* is obtained for some reason, there is a 0.500 probability of PRR− occurring, representing a risk. In other words, managers who invest in *PRP* will always obtain a favorable result associated with *PRR*; that is, applying the *PUS*, *SLP*, and *SUN* tools favor the execution and implementation of *UPL*, *INMI*, and *FLR*.

When performing the sensitivity analysis of the relationship between *MAC* and *QUC*, the high relationship between these variables is again observed since if *MAC+* occurs, then there is a 0.378 probability of *QUC+* occurring. However, a null probability of *QUC−* occurring. However, if *MAC−* occurs, then *QUC+* has only a 0.026 probability of occurring, but under those same *MAC−* conditions, *QUC−* can occur with a probability of 0.590. In other words, *MAC* guarantees the success of *QUC* in productive systems; that is, the implementation of LM tools associated with *CEL*, *SMED*, and *TPM* is a guarantee to obtain *QUC*. Thus, managers who invest in maintenance programs and rapid changeovers on production lines can facilitate quality assurance activities.

It is essential to mention that *QUC* should be planned using LM tools in *PRP*, as H$_5$ indicates and the sensitivity analysis demonstrates, since if *PRP+* occurs, then there is a 0.400 probability of *QUC+* occurring. However, if those same *PRP+* conditions hold, only a 0.033 chance of *QUC−* occurs. However, if *MAC−* is present, there is a zero conditional probability of *QUC+* occurring, but if the *MAC−* conditions hold, then there is a 0.500 chance of *QUC−* occurring, representing a risk. In other words, the implementation of LM tools or practices, such as *UPL*, *INMI*, and *FLR*, guarantee that *QUC* will be obtained in the production lines.

Finally, in the relationship between *PRR* and *QUC*, H$_6$ shows the importance of implementing planned activities in the production lines. It is observed that if *PRR+* is present, then there is a probability of 0.533 that *QUC+* will occur, but if *PRR+* maintains these conditions, then *QUC−* cannot occur. In other words, the investments made to implement *UPL*, *INMI*, and *FLR* always guarantee quality management. However, if *PRR−* occurs, then there is only a 0.027 probability of *QUC+* occurring, and if *PRR−* maintains those conditions, then there is a 0.595 probability of *QUC−* occurring.

Annex: Additional Data for Support Validation

T ratios for path coefficients

	PRP	*MAC*	*PRR*
QUC	2.655	5.380	5.722
PRP		8.486	
PRR	3.398	10.009	

Confidence intervals for path coefficients

	PRP		MAC		PRR	
	LCL	UCL	LCL	UCL	LCL	UCL
SLP	0.045	0.301	0.217	0.465	0.237	0.485
UPL			0.400	0.640		
QUC	0.093	0.347	0.486	0.722		

T ratios for loadings

Item	QUC	PRP	MAC	PRR
QUC01	12.721			
QUC02	12.781			
QUC03	14.542			
QUC04	11.894			
QUC07	13.013			
QUC08	11.462			
lv-PUS		14.476		
lv-SLP		11.995		
lv-SUN		13.775		
lv-CEL			13.410	
lv-TPM			13.840	
lv-SMED			14.362	
lv-INMI				11.309
lv-UPL				14.365
lv-FLR				13.914

Note The prefix "lv" means that it refers to a latent variable. Remember that this is a second-order model

Confidence intervals for loadings

	QUC		PRP		MAC		PPR	
	LCL	UCL	LCL	UCL	LCL	UCL	LCL	UCL
QUC01	0.633	0.863						
QUC02	0.636	0.866						
QUC03	0.727	0.954						
QUC04	0.589	0.821						
QUC07	0.648	0.878						
QUC08	0.565	0.799						

(continued)

(continued)

	QUC		PRP		MAC		PPR	
	LCL	UCL	LCL	UCL	LCL	UCL	LCL	UCL
lv-PUS			0.724	0.951				
lv-SLP			0.594	0.826				
lv-SUN			0.688	0.916				
lv-CEL					0.669	0.898		
lv-TPM					0.691	0.919		
lv-SMED					0.718	0.945		
lv-INMI							0.557	0.791
lv-UPL							0.718	0.945
lv-FLR							0.695	0.923

Note The prefix "lv" means that it refers to a latent variable. Remember that this is a second-order model

PLSc reliabilities (Dijkstra's rho_a's)

QUC	PRP	MAC	PRR
0.856	0.741	0.742	0.661

Additional indices (indicator corr. matrix fit)

- Standardized root mean squared residual (SRMR) = 0.058, acceptable if ≤ 0.1
- Standardized mean absolute residual (SMAR) = 0.048, acceptable if ≤ 0.1
- Standardized chi-squared with 104 degrees of freedom (SChS) = 2.377, $P < 0.001$
- Standardized threshold difference count ratio (STDCR) = 0.905, acceptable if $\geq$ 0.7, ideally = 1
- Standardized threshold difference sum ratio (STDSR) = 0.863, acceptable if $\geq$ 0.7, ideally = 1.

Additional reliability coefficients

	QUC	PRP	MAC	PRR
Dijkstra's PLSc reliability	0.856	0.741	0.742	0.661
True composite reliability	0.885	0.827	0.848	0.817
Factor reliability	0.885	0.827	0.848	0.817

Correlations among l.vs. with sq. rts. of AVEs

	QUC	PRP	MAC	PRR
QUC	0.750	0.502	0.711	0.712
PRP	0.502	0.785	0.501	0.503
MAC	0.711	0.501	0.807	0.741
PRR	0.712	0.503	0.741	0.775

Full collinearity VIFs

QUC	PRP	MAC	PRR
2.460	1.448	2.669	2.681

HTMT ratios (good if < 0.90, best if < 0.85)

	QUC	PRP	MAC
PRP	0.646 ($p < 0.001$)		
MAC	0.890 ($p = 0.010$)	0.702 ($p < 0.001$)	
PRR	0.848 ($p = 0.010$)	0.772 ($p < 0.001$)	0.901 ($p < 0.015$)

References

A.N. Adnan, N.A. Arbaai, A. Ismail, Improvement of overall efficiency of production line by using line balancing. ARPN J. Eng. Appl. Sci. **11**(12), 7752–7758 (2016)

M.F. Ahmad, S.F. Zamri, Y. Ngadiman, C.S. Wei, N.A. Hamid, A.N.A. Ahmad, M.N.M. Nawi, N.A.A. Rahman, The impact of total productive maintenance (TPM) as mediator between total quality management (TQM) and business performance. Int. J. Supply Chain Manag. **8**(1), 767–771 (2019)

J.G. Alfonso, M. Mara, Rewards for continuous training: a learning organisation perspective. Ind. Commer. Train. **48**(5), 257–264 (2016). https://doi.org/10.1108/ICT-11-2015-0076

M.J. Anzanello, F.S. Fogliatto, Selecting the best clustering variables for grouping mass-customized products involving workers' learning. Int. J. Prod. Econ. **130**(2), 268–276 (2011). https://doi.org/10.1016/j.ijpe.2011.01.009

A. Awasthi, Supplier quality evaluation using a fuzzy multi criteria decision making approach, in *Studies in Fuzziness and Soft Computing*, vol. 319 (2015). https://doi.org/10.1007/978-3-319-12883-2_7

A. Badiane, S. Nadeau, J.P. Kenné, V. Polotski, Optimizing production while reducing machinery lockout/tagout circumvention possibilities. J. Qual. Maint. Eng. **22**(2), 188–201 (2016). https://doi.org/10.1108/JQME-04-2014-0015

J. Böllhoff, J. Metternich, N. Frick, M. Kruczek, Evaluation of the human error probability in cellular manufacturing. Procedia CIRP 218–223 (2016). https://doi.org/10.1016/j.procir.2016.07.080

S. Chen, M. Tseng, Procuring customized products: integrating contracting with co-desing, in *POMS 18th Annual Conference*, Dallas, TX, 4–7 May 2007

C.-H. Chung, Quality control sampling plans under zero inventories: an alternative method. Prod. Inventory Manag. **28**(2), 37–42 (1987)

D.E. Crews, M.J. Russ, The impact of individual differences on multitasking ability. Int. J. Product. Perform. Manag. **69**(6), 1301–1319 (2020). https://doi.org/10.1108/IJPPM-04-2019-0191

V. Deac, Impact of industrial maintenance in quality assurance IV. Total productive maintenance (TPM)—guarantee of the quality (I). Qual. Access Success **11**(12), 23–27 (2010)

V. Deac, Impact of industrial maintenance in quality assurance IV. Total productive maintenance (TPM)—guarantee of the quality (II). Qual. Access Success **12**(1), 40–44 (2011)

J. Díaz-Reza, J. García-Alcaraz, L. Avelar-Sosa, J. Mendoza-Fong, J. Sáenz Diez-Muro, J. Blanco-Fernández, The role of managerial commitment and TPM implementation strategies in productivity benefits. Appl. Sci. **8**(7), 1153 (2018). https://doi.org/10.3390/app8071153

R. Ernst, J. Guerrero, A. Roshwalb, A quality control approach for monitoring inventory stock levels. J. Oper. Res. Soc. **44**(11), 1115–1127 (1993). https://doi.org/10.1057/jors.1993.184

J.W. Fowler, S.-H. Kim, D.L. Shunk, Design for customer responsiveness: decision support system for push–pull supply chains with multiple demand fulfillment points. Decis. Support Syst. **123**, 113071 (2019). https://doi.org/10.1016/j.dss.2019.113071

J.L. García-Alcaraz, G. Alor-Hernández, C. Sánchez-Ramírez, E. Jiménez-Macías, J. Blanco-Fernández, J.I. Latorre-Biel, Mediating role of the six sigma implementation strategy and investment in human resources in economic success and sustainability. Sustainability (Switzerland) **10**(6) (2018). https://doi.org/10.3390/su10061828

J.L. García-Alcaraz, V. Martínez-Loya, J.R. Díaz-Reza, J. Blanco-Fernández, E. Jiménez-Macías, A.J. Gil-López, Effect of ICT integration on SC flexibility, agility and company' performance: the Mexican maquiladora experience. Wireless Networks, **26**(7), 4805-4818 (2020). https://doi.org/10.1007/s11276-019-02068-6

P. Gasparini, R. Bertani, F. De Natale, L. Di Cosmo, E. Pompei, Quality control procedures in the Italian national forest inventory. J. Environ. Monit. **11**(4), 761–768 (2009). https://doi.org/10.1039/b818164k

G. Gathen, What can SMED do for you? Ind. Maint. Plant Oper. **65**(7), 10–12 (2004)

A.J. Gil, J.L. García-Alcaraz, M. Mataveli, The effect of learning culture on training transfer: empirical evidence in Spanish teachers. Int. J. Hum. Resour. Manag. 1–24 (2018). https://doi.org/10.1080/09585192.2018.1505763

R.L. Hannan, G.P. McPhee, A.H. Newman, I.D. Tafkov, S.J. Kachelmeier, The informativeness of relative performance information and its effect on effort allocation in a multitask environment. Contemp. Account. Res. **36**(3), 1607–1633 (2019). https://doi.org/10.1111/1911-3846.12482

A. Hashemoghli, I. Mahdavi, A. Tajdin, A novel robust possibilistic cellular manufacturing model considering worker skill and product quality. Sci. Iran. **26**(1E), 538–556 (2019). https://doi.org/10.24200/sci.2018.4948.1002

L.A. Ilyina, M.A. Brazhnikov, I.V. Khorina, Formation of a complex system of indicators for evaluation of smooth production flow. Contrib. Econ. (2017). https://doi.org/10.1007/978-3-319-45462-7_48

A.A. Karam, M. Liviu, V. Cristina, H. Radu, The contribution of lean manufacturing tools to changeover time decrease in the pharmaceutical industry. A SMED project. Procedia Manuf. 886–892 (2018). https://doi.org/10.1016/j.promfg.2018.03.125

B. Kennedy, Accelerating productivity: applying multitasking strategies to boost output and smooth production flow. Cut. Tool Eng. **55**(7), 30 (2003)

H.J. Kim, T.E. Lee, Scheduling of cluster tools with ready time constraints for small lot production, in *IEEE International Conference on Automation Science and Engineering* (2011), pp. 96–101. https://doi.org/10.1109/CASE.2011.6042518

M. Kolosowski, J. Duda, J. Tomasiak, Statistical process control in conditions of piece and small lot production, in *Annals of DAAAM and Proceedings of the International DAAAM Symposium* (2015), pp. 147–155. https://doi.org/10.2507/26th.daaam.proceedings.021

C.J. König, L. Oberacher, M. Kleinmann, Personal and situational determinants of multitasking at work. J. Pers. Psychol. **9**(2), 99–103 (2010). https://doi.org/10.1027/1866-5888/a000008

S. Kristoffersen, Production, takt and lead time variation: production planning and tardiness probabilities in small-lot production, in *2014 International Conference on Collaboration Technologies and Systems, CTS 2014* (2014), pp. 236–243. https://doi.org/10.1109/CTS.2014.6867570

C.M. Lin, L.F. Chen, C.T. Su, Supplier quality management in TFT-LCD new product development: a case study. J. Qual. **23**(6), 351–374 (2016). https://doi.org/10.6220/joq.2016.23(6).01

Y.M. Méndez-Vázquez, D.A. Nembhard, Worker-cell assignment: the impact of organizational factors on performance in cellular manufacturing systems. Comput. Ind. Eng. **127**, 1101–1114 (2019). https://doi.org/10.1016/j.cie.2018.11.050

G. Nawanir, Y. Fernando, K.T. Lim, The complementarity of lean manufacturing practices with importance-performance analysis: how does it leverage inventory performance? Int. J. Serv. Oper. Manag. **39**(2), 212–234 (2021). https://doi.org/10.1504/IJSOM.2021.115451

J. Nelson, Pull versus push: lessons from lean manufacturing, Chap. 3, in *Becoming a Lean Library* (Chandos Publishing, 2016), pp. 29–49. https://doi.org/10.1016/B978-1-84334-779-8.00003-3

D. Ojha, J. Shockley, C. Acharya, Supply chain organizational infrastructure for promoting entrepreneurial emphasis and innovativeness: the role of trust and learning. Int. J. Prod. Econ. **179**, 212–227 (2016). https://doi.org/10.1016/j.ijpe.2016.06.011

J.C. Osorio Gómez, J.L. García Alcaraz, D.F. Manotas Duque, AHP TOPSIS for supplier selection considering the risk of quality. Espacios **39**(16) (2018)

K.T. Reynolds, Cellular manufacturing & the concept of total quality. Comput. Ind. Eng. **35**(1–2), 89–92 (1998). https://doi.org/10.1016/S0360-8352(98)00027-8

C.L. Rus, S. Chirică, L. Rațiu, A. Băban, Learning organization and social responsibility in Romanian higher education institutions. Procedia Soc. Behav. Sci. **142**, 146–153 (2014). https://doi.org/10.1016/j.sbspro.2014.07.628

H. Salimian, M. Rashidirad, E. Soltani, Supplier quality management and performance: the effect of supply chain oriented culture. Prod. Plan. Control 1–17 (2020). https://doi.org/10.1080/09537287.2020.1777478

K.R. Sanderson, V. Bruk-Lee, C. Viswesvaran, S. Gutierrez, T. Kantrowitz, Multitasking: do preference and ability interact to predict performance at work? J. Occup. Organ. Psychol. **86**(4), 556–563 (2013). https://doi.org/10.1111/joop.12025

M.M. Savino, S. Apolloni, Y. Ouzrout, Product quality pointers for small lots production: a new driver for quality management systems. Int. J. Prod. Dev. **5**(1–2), 199–211 (2008). https://doi.org/10.1504/IJPD.2008.016378

V. Schindlerová, I. Šajdlerová, V. Michalčík, J. Nevima, L. Krejčí, Potential of using TPM to increase the efficiency of production processes. Teh. Vjesn. **27**(3), 737–743 (2020). https://doi.org/10.17559/TV-20190328130749

V.B. Schramm, L.P.B. Cabral, F. Schramm, Approaches for supporting sustainable supplier selection—a literature review. J. Clean. Prod. **273**, 123089 (2020). https://doi.org/10.1016/j.jclepro.2020.123089

H. Seifoddini, M. Djassemi, The threshold value of a quality index for formation of cellular manufacturing systems. Int. J. Prod. Res. **34**(12), 3401–3416 (1996). https://doi.org/10.1080/00207549608905097

S. Sharma, A. Singh, Inventory minimisation by service level optimisation for increased freshness and availability to end consumer in a multi echelon system: FMCG case. Int. J. Appl. Manag. Sci. **4**(2), 165–188 (2012). https://doi.org/10.1504/IJAMS.2012.046231

K. Singh, I.P.S. Ahuja, Assessing the business performance measurements for transfusion of TQM and TPM initiatives in the Indian manufacturing industry. Int. J. Technol. Policy Manage. **14**(1), 44–82 (2014). https://doi.org/10.1504/IJTPM.2014.058733

I. Suija-Markova, L. Briede, E. Gaile-Sarkane, I. Ozoliņa-Ozola, Multitasking in knowledge intensive business services. Emerg. Sci. J. **4**(4), 305–318 (2020). https://doi.org/10.28991/esj-2020-01233

M. Tekin, M. Arslandere, M. Etlioğlu, Ö. Koyuncuoğlu, E. Tekin, An application of SMED and Jidoka in lean production, in *Proceedings of the International Symposium for Production Research 2018*, ed. by N.M. Durakbasa, M.G. Gencyilmaz (Springer International Publishing, Cham, 2019), pp. 530–545

E. Turcu, V. Bendic, C. Mohora, D. Tilină, A.M. Niță, Improvement of maintenance process by practical approach of total productive maintenance (TPM) in the view of quality management, in *Proceedings of the 28th International Business Information Management Association Conference—Vision 2020: Innovation Management, Development Sustainability, and Competitive Economic Growth* (2016), pp. 4187–4199

S. Valipour Parkouhi, A. Safaei Ghadikolaei, H. Fallah Lajimi, Resilient supplier selection and segmentation in grey environment. J. Clean. Prod. **207**, 1123–1137 (2019). https://doi.org/10.1016/j.jclepro.2018.10.007

S. Vijay, M. Gomathi Prabha, Work standardization and line balancing in a windmill gearbox manufacturing cell: a case study. Mater. Today Proc. (2020). https://doi.org/10.1016/j.matpr.2020.08.584

Z.T. Xiang, C.J. Feng, Implementing total productive maintenance in a manufacturing small or medium-sized enterprise. J. Ind. Eng. Manag. **14**(2), 152–175 (2021). https://doi.org/10.3926/jiem.3286

K. Yazıcı, S.H. Gökler, S. Boran, An integrated SMED-fuzzy FMEA model for reducing setup time. J. Intell. Manuf. (2020). https://doi.org/10.1007/s10845-020-01675-x

R. Yousef, N.K. La Scola, K.D. Walsh, T.C.L. Da Alves, Supplier quality management inside and outside the construction industry. EMJ Eng. Manag. J. **27**(1), 11–22 (2015). https://doi.org/10.1080/10429247.2015.11432032

Z.F. Zhang, D. Janet, An entropy-based approach for measuring the information quantity of small lots production in a flow shop. Zidonghua Xuebao/Acta Autom. Sin. **46**(10), 2221–2228 (2020). https://doi.org/10.16383/j.aas.c180479

Q. Zhang, W. Tang, J. Zhang, Green supply chain performance with cost learning and operational inefficiency effects. J. Clean. Prod. **112**, 3267–3284 (2016). https://doi.org/10.1016/j.jclepro.2015.10.069

M. Zhang, H. Guo, B. Huo, X. Zhao, J. Huang, Linking supply chain quality integration with mass customization and product modularity. Int. J. Prod. Econ. **207**, 227–235 (2019). https://doi.org/10.1016/j.ijpe.2017.01.011

Q. Zou, H.-Y. Feng, Push-pull direct modeling of solid CAD models. Adv. Eng. Softw. **127**, 59–69 (2019). https://doi.org/10.1016/j.advengsoft.2018.10.003

Printed in the United States
by Baker & Taylor Publisher Services